I0038505

New Hybrid Quantum Big_Bang-Megaverse_Driven Universe with a Finite Big Bang and an Increasing Hubble Constant

SECOND EDITION of *Quantum Big Bang Cosmology*
Directly Links the Big Bang with the Present Expansion
An Initial Quantum Big Bang Metastate with No Infinities
A Megaverse Energy Influx Fueling Hubble Growth
A Big Dip in the Hubble Parameter

Stephen Blaha Ph. D.
Blaha Research

To strive, to seek, to find, and not to yield.
Ulysses – A. Tennyson

Pingree-Hill Publishing
MMXIX

Copyright © 2019 by Stephen Blaha. All Rights Reserved.

This document is protected under copyright laws and international copyright conventions. No part of this book may be reproduced, stored in a retrieval system, or transmitted by any means in any form, electronic, mechanical, photocopying, recording, or as a rewritten passage(s), or otherwise, without the express prior written permission of Blaha Research. For additional information send an email to the author at sblaha777@yahoo.com.

ISBN: 978-1-7328245-3-9

This document is provided "as is" without a warranty of any kind, either implied or expressed, including, but not limited to, implied warranties of fitness for a particular purpose, merchantability, or non-infringement. This document may contain typographic errors, technical inaccuracies, and may not describe recent developments. This book is printed on acid free paper.

Cover: The cover depicts space and a miniature of the First Edition of this book..

Rev. 00/00/01   June 8, 2019

To My Wife Margaret

Some Other Books by Stephen Blaha

All the Megaverse! Starships Exploring the Endless Universes of the Cosmos using the Baryonic Force (Blaha Research, Auburn, NH, 2014)

SuperCivilizations: Civilizations as Superorganisms (McMann-Fisher Publishing, Auburn, NH, 2010)

All the Universe! Faster Than Light Tachyon Quark Starships & Particle Accelerators with the LHC as a Prototype Starship Drive Scientific Edition (Pingree-Hill Publishing, Auburn, NH, 2011).

Cosmos Creation: The Unified SuperStandard Model, Volume 2, SECOND EDITION (Pingree Hill Publishing, Auburn, NH, 2018).

Immortal Eye: God Theory: Second Edition (Pingree Hill Publishing, Auburn, NH, 2018).

Unification of God Theory and Unified SuperStandard Model THIRD EDITION (Pingree Hill Publishing, Auburn, NH, 2018).

Calculation of: QED α = 1/137, and Other Coupling Constants of theUnified SuperStandard Theory (Pingree Hill Publishing, Auburn, NH, 2019).

Coupling Constants of the Unified SuperStandard Theory SECOND EDITION: We Find the Fine Structure Constant 1/137.0359801, and so: OUR UNIVERSE AND LIFE! Also a Universal Eigenvalue Function for all Known Interactions, And Running Coupling Constants to all Perturbative Orders (Pingree Hill Publishing, Auburn, NH, 2019).

Available on Amazon.com, bn.com Amazon.co.uk and other international web sites as well as at better bookstores (through Ingram Distributors).

CONTENTS

SECTION 3: A NEW PHENOMENOLOGICAL THEORY OF THE HUBBLE PARAMETER 165

FIGURES and TABLES

INTRODUCTION

This book provides a combined theory composed of the theories of Blaha (2004) and (2017c), and a new phenomenological theory of the growth of the universe.

It proposes a quantization of the Big Bang that avoids the infinities normally associated with the Big Bang at zero time. Just as quantization stabilizes an atom to avoid the infinities of the collapse of an atom's electrons into its nucleus, the quantized Big Bang theory stabilizes the Big Bang as an ultra-short-lived metastate without infinities.

After developing this part of the theory the book then proceeds to consider the increasing growth rate of the universe as shown in the changing values of the Hubbble "Constant." It suggests that the author's earlier theory of a larger universe, called the Megaverse, enables the universe to grow by absorbing Megaverse mass-energy through a surface tension force. A detailed study shows that a Megaverse energy influx is possibly a source for the recent growth of the universe.

Lastly the author develops a phenomenological theory of the growth of the universe which has several remarkable features. It dovetails nicely with the quantized Big Bang model. It shows the accelerated growth of the Big Bang metastate decreases dramatically as the growth pressures in the Big Bang metastate decline. It shows that the universe actually contracts as the "gaseous" radiation dominated stage transitions to a "more compact" matter dominated stage. This decline is called the "Big Dip" in the universe's growth in the book.

It then accounts for the growth of the Hubble Parameter H(t) due to an energy influx that presumably comes from the Megaverse. Thus the measured growth in the Hubble Constant from 67.8 to the low 70's is explained.

Lastly it suggests that the success of the phenomenological model is possibly due to a self-similarity in the universe's growth based on the Renormalization group.

SECTION 1: Megaverse - THE SPACE OF SPACES

This section describes a space of more than four dimensions, within which our universes as well as "many" other universes reside. We picture this space, which we call the *Megaverse* to distinguish it from other types of spaces that are called Multiverses, as a container for many coexisting universes. Universes are distributed through the Megaverse in a manner analogous to the distribution of galaxies within our universe. Universes are dynamic: they are created; they are destroyed, they combine to form larger universes; they "interact" we each other rather like elementary particles interact.

In this section we describe the general features of the Megaverse. The material in this section has largely appeared scattered throughout earlier books by the author written since 2004. The reader familiar with this earlier work may either skim or skip this section.

1. Evidence for Entities Beyond Our Universe

1.1 Theoretical Evidence for Other Universes and External Matter

Why are we not content with one universe given its enormous size and variety? It appears that there are important theoretical reasons, and some important experimental observations, that suggest that there is more than our universe 'out there.' The external entities are likely to be other universes but external "clumps" of mass-energy are not excluded.

In this chapter[1] we will discuss theoretical reasons and experimental suggestions of a larger space—that we call the *Megaverse*—that contains our universe and, most likely, other universes. The existence of a Megaverse resolves important theoretical issues and may resolve some important astronomical puzzles that have appeared in recent years.

The theoretical issues, which have been subjects of discussion for some time, are:

1. The need for an external 'clock' to measure 'time' knowing that it is to some extent relative and local.
2. The need for a 'quantum observer' to complete the understanding of quantum gravity as described by the Wheeler-DeWitt equation and other efforts to develop a quantum gravity.
3. The need for other universes to provide theoretical measuring platforms for quantities beyond the charge and mass-energy of the universe. We think here of the other quantum numbers of particles and particle number operators such as Baryon number.
4. The need for an ultimate source of mass and inertia in our universe.

---

[1] Most of this chapter appears in Blaha (2015a) and in earlier books by the author.

In Blaha (2015a), and earlier books, we have suggested that there are weighty reasons to believe that other universes exist.[2] The existence of other universes solves these problems.

These problems generally have a source in Quantum Gravity and the interpretation of the Wheeler-DeWitt equation in particular. See Blaha (2017c) and (2018e) for discussions of the Wheeler-DeWitt equation and its implications. We now consider the above issues.

1.1.1 Universe Clocks

Asynchronous Logic provides the equivalent of a clock for the synchronization of processes within large electrical systems such as VLSI chips. Similarly there is a need for a universal clock for our universe. As DeWitt[3] points out in his studies of quantum gravity,

'"The variables … [of the quantized Friedmann model] because of their lack of hermiticity, are not rigorously observable and hence cannot yield a measure of proper time which is valid under all circumstances. … . It is for this reason that we may say that "time" is only a phenomenological concept … If the principle of general covariance is truly valid then the quantum mechanics of everyday usage with its dependence on the Schrödinger equations … is only a phenomenological theory. For the only "time" which a covariant theory can admit is an intrinsic time defined by the contents of the universe itself. Any intrinsically defined time is necessarily non-Hermitean, which is equivalent to saying that there exists no clock, whether geometrical or material, which can yield a measure of time which is operationally valid under *all* circumstances, and hence there exists no operational method for determining the Schrödinger state function with arbitrarily high precision."

---

[2] In Blaha (2013a), before the Higgs particle was discovered at CERN we suggested an alternate mechanism was possible if a sister universe existed (making the existence of other universes a reasonable possibility. The Higgs discovery makes the sister universe mechanism unlikely.

[3] DeWitt, B. S., Phys. Rev. **160**, 1113 (1987).

The lack of a clock within our universe invalidates quantum mechanics in principle and Quantum Gravity in particular. DeWitt concludes, "Thus [quantum gravity] will say nothing about time unless a clock to measure time is provided." And it appears absolutely necessary.

Unruh[4] also has an issue with the source of time:

"One of the key problems is that of time. We see and experience the world in terms of time. We see things grow, develop, and change. However, time does not enter into the Euclidean formulation of quantum gravity directly. In the usual Hamiltonian formulation, the Hamiltonian for quantum gravity is made up of densities which are the generators, not only of spatial coordinate transformations, but also of temporal coordinate transformations. The content of four of Einstein's equations is that some generators are zero. Thus all wave functions are invariant under all spatial and all temporal coordinate transformations. There is nothing in the wave function or the amplitudes which refers to the coordinate t, or the corresponding points of the manifold in any way. How then do we recover the indubitable and ubiquitous experience we have of time? The standard answer is that our experience of time is actually an experience of different correlations between physical quantities in the world. Time is replaced by the readings of clocks. I know that time has changed, not through any direct experience with time, but because the hands of my watch have changed.

Although the implementation of this idea is actually extremely difficult in practice, and although I personally believe that one should formulate one's quantum theory of gravity so as to contain time explicitly, let us nevertheless pursue the consequences of this idea of time as defined internally, as the "reading" of a dynamic variable. For an observer inside the theory, his "time" is not the coordinate; rather his time is some one of the given dynamic variables of the theory: y or P. Thus although the coupling to the baby universes via the effective action S is independent of the coordinates t or x, that does not mean that the observer inside the theory will experience the interactions as being independent of time. For him and/or her, time is one of the dynamic variables and so it can depend on the various dynamic variables of the theory,

---

[4] Unruh, W. G., Phys. Rev. D **40**, 1053 (1989).

even if it does not depend on the time coordinate t. In general one would expect the observer to see what looks to him like a time-dependent interaction with the baby universes. At one time, some one of the baby universes may couple strongly to the large universe, while at some other time, another of the baby universes will couple more strongly."

In Blaha (2015a) and earlier books, we suggested the existence of other universes provides a 'clock mechanism' in principle for our universe. And being universes themselves, these other universes are excellent clocks. DeWitt points out,

"Because every clock has a "one-sided" energy spectrum, its ultimate accuracy must necessarily be inversely proportional to its rest mass. When the whole universe is cast in the role of a clock, the concept of time can of course be made fantastically accurate (at least in principle) … "

Setting a mass scale using other universes, also sets[5] a time scale and resolves the issue of a clock for our universe. *In principle the existence of other universes validates the role of time in the Copenhagen interpretation of Quantum Mechanics.*

1.1.2 Quantum Observer

Attempts to create a quantum gravity theory have to confront the need for an *Observer* in any quantum theory within the context of the Copenhagen interpretation. DeWitt points out,

"The Copenhagen view depends on the assumed a priori existence of a classical level to which all questions of observation may ultimately be referred. Here, however, the whole universe is the object of inspection; there is no classical vantage point, and hence the interpretation question must be re-argued from the beginning. While we do not wish to stress this point unduly, since, after all, the Friedmann model ignores the

---

[5] For example the Planck time value is set by the Planck mass.

vast complexities of the real universe, it is nevertheless clear that the quantum theory of space-time must ultimately force a deviation from the traditional Copenhagen doctrine."

And Unruh states

"One of the key features in the interpretation of such transition amplitudes, or wave functions, is the idea that we, as observers are also a part of the Universe as a whole. We, as physical observers, must be describable from within the theory and not as observers external to the theory as in usual quantum mechanics. In usual quantum mechanics, the interpretation is usually given in terms of observers that are outside of the theory. There one makes a split, with the quantum world at one side of the split, and the observer on the other. von Neumann argued that the predictions of quantum mechanics, at least under certain assumptions, are independent of the exact location of that split, but Bohr argued adamantly for the necessity of such a split (classical observers and quantum world). *There is a great difficulty in setting up such a split for physical observers contained within and influenced by a quantum universe,* [italics added] and for the Universe as a whole, especially including gravity, one cannot argue that the predictions will be independent of where one puts the split. Since all energies interact gravitationally, and our observations are surely energetic phenomenon, the treatment of the energetics of observation as classical would lead to different predictions than if they were treated quantum mechanically. One is therefore forced to devise an interpretation of quantum mechanics in which the observer is part of the quantum system, rather than outside the quantum system.

This means that the interpretation of these transition amplitudes becomes somewhat non-intuitive. One must ask what the system looks like from within, from the viewpoint of an observer who is part of that world, rather than being able to interpret them directly in terms of probabilities for observations made by an external observer."

While the *Observer* question is addressed by a number of authors, the proposed answers are not entirely convincing. *The existence of other universes provides macroscopic Quantum Observers for our universe.* And our universe acts as a macroscopic quantum observer for other universes. Thus the quantum observer issue is resolved within the Megaverse of universes.

These considerations lead us to view the existence of other universes as a critical solution to the above problems.

1.1.3 The Higgs Mechanism is Explainable by Extra Dimensions

The Higgs Mechanism 'explains' (generates) fermion and boson masses. However the Higgs potential contains a quadratic term with a constant with the dimensions of [mass]. In a sense the Higgs Mechanism trades one mass for another. From where do the Higgs potentials' masses come?

A further explanation is needed is to determine the origin of the "dimensionful" mass terms in the Higgs' particle equations themselves. At present little if any thought has been given to the origin of these terms. We suggested that, excluding a "deus ex machina" source, the only known way to generate these mass terms in the Higgs' equations is through the separation of equations technique of differential equations. This technique requires additional parameters which can only be the coordinates of *extra unknown dimensions*. The best example of the generation of mass terms appears in the Schwarzschild solution of General Relativity where a separation constant, often denoted M, appears that has the dimension of [mass].

Thus extra space-time dimensions would resolve the origin of Higgs potentials' masses. Given extra dimensions it is reasonable to expect that these extra dimensions contain universes. Thus the Megaverse!

1.1.4 Possible Accretion of Megaverse Matter to Fuel Expansion of Our Universe

If matter is distributed outside of universes in the Megaverse, and if this matter can be accreted to universes by gravitational attraction, then the apparent increasing expansion of our universe may be due to this accretion. In chapter 14 of Blaha (2017c) we presented a model in which this possibility is realized. If true, then we would have tangible evidence of the existence of other universes in the Megaverse.

1.1.5 Asynchronous Logic is a Requirement of Universes

By establishing Asynchronous Logic principles[6] as the basis for the existence of universes and for setting the number of dimensions in each universe – four; and establishing the basis of fermion particles as *qubes*—we have found deeper principles of organization for the foundations of physics. The principles built on this foundation serve to enable the coordination of complex physical processes.

Usually we look at particle processes primarily from a space-time perspective: particles collide and produce new particles. We primarily think of the incoming and outgoing particles in a collision. However, considering the set of fundamental particles – and particle transforming interactions in themselves, while neglecting space-time and momentum considerations, leads us to view particles as constituting an alphabet and to view their interactions as a type of computer grammar.[7] Then the Asynchronicity Principles enable us to bring in space-time in a way that gives us the maximum complexity with the most minimal assumptions. As Leibniz[8] points out our universe has maximal complexity with minimal assumptions.

1.1.6 The Meaning of Total Quantities of a Universe

The 'external' properties of a universe are normally questioned—for the simple reason that it is assumed that there is no 'outside' of our universe. For example, Misner (1973) asserts:[9]

'There is no such thing as "the energy (or angular momentum, or charge) of a closed universe," according to general relativity, and this for a simple reason. To weigh something one needs a platform on which to stand to do the weighing."

---

[6] The basis of this section is described in detail in Blaha (2015a). That book places Physics within a logical framework that is a possible deeper ground for fundamental Physics theory.
[7] This conceptual approach was first described in Blaha (1998) who went on to characterize our universe as one enormous word evolving in time.
[8] See Rescher (1967).
[9] Pp. 457 - 458.

Misner et al presumes no such platform exists. If there is but one closed universe as most physicists currently believe then one cannot measure any totals of a closed universe such as ours. Yet if we take a more general view that our universe is only one of many then it becomes possible to measure total mass, charge, angular momentum, baryon number, and many other quantities of interest. Indeed, the existence of other universes (within the encompassing Megaverse) opens the door to an understanding of time, mass, energy, and all the other quantities necessary to develop a dynamical theory of universes.

Our new 'rotations of interactions' formalism (described in previous books) enables us to rotate measurable quantities. These quantities (quantum numbers) furnish a set of totals for our universe such as baryon numbers (normal and Dark), lepton numbers, angular momentum and so on that characterize our universe.

Later we will also see that one can then treat universes as 'particles', and develop 'universe dynamics', which might explain knotty problems such as the Big Bang and its precursor (if any). We will do this in subsequent chapters after first considering the possible structure of universes in general in the Megaverse.

1.2 Possible Experimental Evidence for the Megaverse

At first glance it would seem impossible to produce evidence for the existence of other universes. However there are subtle means by which we can 'sense' experimentally 'nearby' universes should they exist. The mechanism would appear to be gravitational effects exerted on objects within our universe by unseen nearby objects of enormous mass. Currently there appears to be three experimental suggestions for the existence of 'nearby' universes and one theoretical argument based on an influx of mass-energy from the Megaverse that may support an understanding of the expansion of our universe.

1.2.1 Great Attractors

One potential support is the discovery of the Great Attractor (at the center of the Laniakea Galaxy Supercluster), and the more massive Shapley Attractor (centered in the

Shapley Supercluster)[10]. These attractors contain massive numbers of galaxies and are drawing galaxies over a distance of millions of light years towards them.

If another universe(s) is 'near' our universe it could act as a 'gravitational magnet' and draw galaxies within our universe towards it to form one or more superclusters which could then act as attractors. Thus attractors might indirectly reveal the presence of other nearby universes—contrary to the expected large scale uniformity of the universe. The only other apparent source of superclusters is chance. Chance seems an unsatisfactory possibility in the present case.

1.2.2 Bright Bumps in Universe Suggesting Collision with Another Universe

A recent study[11] of the residual brightness of parts of the accessible universe found that bright patches appeared if a model of the CMB (Cosmic Microwave Background) with gases, stars and dust was 'subtracted' from the PLANCK map of the entire sky. After the subtraction one would expect only noise spread throughout the sky. However, bright patches were seen in a certain range of frequencies. These anomalies are thought to be a result of our universe colliding with another object – presumably another universe in the Megaverse.

1.2.3 Cold Spot in Universe Suggesting Collision with Another Universe

Another recent study[12] of a huge cold region of the universe spanning billions of light years revealed that this region is not a relatively empty region but rather is similar to in its distribution of galaxies to the rest of the universe. Previous the Cold Spot was considered an area where cosmic microwave background radiation – the leftover Big Bang radiation is weak – making it significantly colder (0.00015C colder) than the average temperature of the universe.

An analysis of 7,000 galaxy redshifts using new high-resolution data has now shown that the Cold Spot is similar to the rest of the universe. The Durham University

---

[10] Tully, R. Brent; Courtois, Helene; Hoffman, Yehuda; Pomarède, Daniel, "The Laniakea Supercluster of galaxies". Nature (4 September 2014). 513 (7516): 71–73; arXiv:1409.0880.
[11] Ranga-Ram Chary, arXiv.org:/1510.00126 (2015).
[12] T. Shanks et al, Durham University (Australia), Monthly Notices of the Royal Astronomical Society, 2016.

group suggested that the Cold Spot might have been caused by a collision between our universe and another Universe. They further suggested that there is only a 1 in 50 chance that it could explain this feature with standard cosmology.

Thus we have another important piece of circumstantial evidence in favor of other universes and thus the Megaverse.

1.2.4 Megaverse Energy-Matter Infusion into Our Universe

In chapter 14 of Blaha (2017c) we presented a model for an influx of mass-energy from the Megaverse to support the Bondi-Gold-Hoyle-Narlikar Steady State Cosmology, which was originally based on the 'continuous creation of mass-energy' by Hoyle and Narlikar. This model explains why the value of Ω makes the universe close to flat. If this model is correct then we would have concrete support for a Megaverse with a low mass-energy density leaking mass-energy into our universe. *More generally, it suggests that universes are surfaces of high mass-energy density in a Megaverse of low mass-energy density – with a ratio of mass-energy densities of the other of 10^{30}.*

1.2.5 Conclusion

We conclude that data is beginning to emerge favoring multiple universes and a physical Megaverse in support of the theoretical justifications presented earlier.

1.3 Historical Trend Toward Larger Space-Time Structures

Looking back through the history of Mankind's view of the universe we see a clear progression to a larger and larger view. Before the 16th century the earth was the universe. In the 16th century Giordano Bruno (and possibly others) suggested that the stars were suns with many worlds circling them. So our view of the universe expanded to include stars. Now we are on the threshold of external universes.

1.4 Possibly Universe Fragments

In addition to other universes the possibility of diffuse mass-energy in the Megaverse appears likely. The possibility of "mini-universes" is also likely.

2. Megaverse Dimension

Having seen strong motivation for the existence of a space of spaces (the Megaverse) we now turn to estimating its features. The determination of the dimensions of the Megaverse can only be described as guesswork unless a sound principle is consistently used to specify the dimensions. In this chapter we will use the known interactions (and the equal number of fermions) in our universe[13] to determine the Megaverse's dimension.[14]

We will assume that interactions have a dual role in fundamental physics: they determine the dynamics of particles, and they act to determine the dimensions of the universes and the Megaverse. The first role is evident within the universe and is therefore obvious from experiment. The second role is external to our universe. It acts to determine the dimensions of the Megaverse.

2.1 Role of Interactions in Determining the Megaverse Dimension D

The motivation for the second role can be discerned from considering a 2-dimensional space, and introducing a simple 1/r potential such as:

$$V = g^2/(x^2 + y^2)^{-\frac{1}{2}}$$

where g is a coupling constant.

One views V as the potential of a force. However the values of V suitably extended to the range [−∞, +∞] can be viewed as a third dimension.

With this alternate perspective in mind we take the subset of Unified SuperStandard Model vector interactions,

---

[13] We use the Unified SuperStandard Theory to determine these quantities. See Blaha (2018e).
[14] Our general motivation for this choice is the relation between space-time and particles: without particles the measurement of distance in space-time is meaningless; without space-time particle dynamics is meaningless. The concepts of space and particles are "yoked" together. SuperSymmetry considerations somewhat support this view.

$$E = [SU(3){\otimes}SU(2){\otimes}U(1){\otimes}SU(2){\otimes}U(1){\otimes}U(4){\otimes}U(4)]^4$$

which has 192 generators, to set the Megaverse dimension to[15]

$$D = 192 \tag{2.1}$$

It is important to note that the fields of the Unified Standard Model extend beyond universe boundaries since each point of a universe is surrounded by an infinity of Megaverse points. This feature is described in detail in the following chapter. Under these circumstances we require all fields be equally describable by the coordinates of the universe *and* Megaverse coordinates. Then continuity requires smooth transitions between values at universe points and Megaverse points in the neighborhood of universe points. Thus fields in our universe are intimately connected to their Megaverse counterparts, partially justifying our determination of the Megaverse dimension above. We provide further support for eq. 2.1 when we consider the 'rotation of symmetries' U(192) Θ-Group. (See Blaha (2018e.)

Lastly, we note for consistency, the Megaverse must have complex-valued coordinates since it encompasses the complex-valued coordinates of our universe (although a more complicated embedding can be visualized.) We opt for simplicity. In addition the metric of the Megaverse must be Euclidean with a flat-space metric tensor limit consisting entirely of –1's along the diagonal (using our conventions). For good reason we view the Megaverse as having at best a very low mass-energy density compared to that of our universe. Thus the Megaverse is essentially flat with 'universe bubbles.'

---

[15] We will use the symbol D to denote the dimension of the Megaverse in the following chapters.

3. The Embedding of a Universe in a Higher Dimension Megaverse – Surface Tension

3.1 Universes as Mass-Energy Islands

In developing the theory of the Megaverse we view universes as islands of mass-energy that maintain their 'integrity' as surfaces due to gravitation. Gravity holds universes together rather like molecular forces within a water droplet hold water molecules together. Molecular attractive forces give droplets cohesion as they (perhaps) descend through the earth's atmosphere. They are the origin of *surface tension* in water.

Similarly gravity holds higher density[16] mass-energy universes together and gives rise to gravity surface tension. In a model presented in Blaha (2017c) for the Big Bang and the expansion of the universe within the Megaverse we found the mass-energy density of the universe was a factor of 10^{30} more than the density in the surrounding Megaverse space.

Thus we have good reason to study the surface tension of universes as a result of gravitational attraction within universes.

3.2 Boundary of a Universe within the Megaverse

In this chapter we describe the embedding of universes within the Megaverse. Much of this chapter appears in several earlier books by the author such as Blaha (2015a).

As stated earlier, we define a universe to be a closed or open surface in the Megaverse with a much higher mass-energy density than the Megaverse.

There are two types of boundaries for a universe embedded in a space of larger dimensions. First there is a boundary of the universe determined by treating the universe

---

[16] Higher density in comparison to the much lower density of the inter-universe space of the Megaverse.

as a surface in the space. Secondly, there is another type of universe boundary defined by the observation that any neighborhood of ever

A point of a universe – not strictly within the universe – has an infinite number of points of the enclosing Megaverse space.[17] Or, every universe point has neighborhoods with Megaverse points within it. Thus *each point of a universe is on a boundary of the universe due to the larger dimensions of the Megaverse space* within which it resides. Fig. 3.1 schematically illustrates these neighborhoods for any universe point for a universe contained within a higher dimensional Megaverse.

3.3 Confinement of Universes due to 'Surface Tension'

We will assume that other universes have the same physics as our universe with the possible differences that they may have differing interaction coupling constants and differing particle masses. As we stated above, every point of a universe in a higher dimensional Megaverse has Megaverse points in any neighborhood of the point (with the exception of neighborhoods strictly within the universe). Thus we confront the question: what keeps mass-energy at points in a universe, or is there leakage from the universe into the Megaverse?

If there is little or no leakage into the Megaverse then, since each point in a universe is part of a Megaverse surface, one can only assume that there is a barrier to movement into the Megaverse. Taking a note from fluid dynamics, and viewing Megaverse space as one 'material' and the universe as a different 'material,'[18] we view the barrier as 'surface tension.'[19] The Megaverse appears to "exert a force" confining the contents of the universe to within itself.[20]

---

[17] Any neighborhood of any point in the universe – with all its points strictly within the universe – has an infinite number of points within the universe. We assume the neighborhood is so small that the curvature of the universe's space can be neglected.

[18] Meaning material with much higher mass-energy density and consequently larger internal gravitational attraction.

[19] See Landau (1987).

[20] Although in actuality it is the universe that holds itself together by gravitation. The surface tension force is the result of the internal gravitation of mass-energy within the universe.

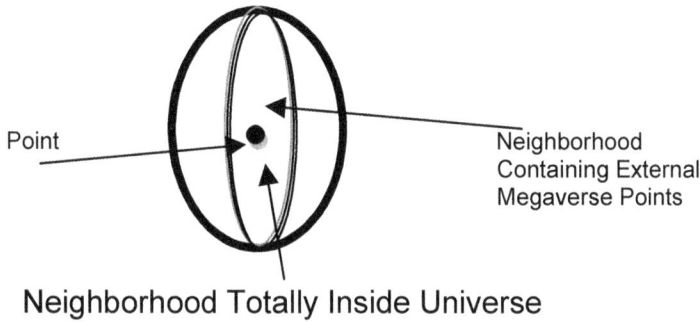

Neighborhood Totally Inside Universe

Figure 3.1. Schematic diagram of a 3-dimensional projection of 'orthogonal' neighborhoods of a point within a universe with one neighborhood strictly within the universe and the other neighborhood containing external Megaverse points in general. The 'orthogonal' circles around the point differentiate between the two types of neighborhoods.

The surface tension[21] of a universe γ satisfies the relation

$$\gamma = W/\Delta A \qquad (3.1)$$

where γ is expressed in erg/cm^2, W is the Work, and ΔA is the Area upon which the work is exerted. The pressure Δp exerted by the surface tension for a 'spherical' surface area is

$$\Delta p = 2\gamma/R \qquad (3.2)$$

where R equals the radius of curvature of the surface. The above equations embody the concept that the surface tension force equals the pressure difference at the surface.

---

[21] A useful analogy: the Megaverse is a pool of water; a universe is a denser oil bubble within it. Surface tension caused by the cohesiveness of the oil molecules in the bubble makes it spherical (confines it to a spherical shape). Similarly a universe (denser than the Megaverse) is 'confined' within the Megaverse.

If the universe is flat then the surface pressure approaches ∞ giving confinement of the fields and particles to the universe:

$$R \rightarrow 0 \quad \text{implies } \Delta p \rightarrow \infty \quad\quad (3.3)$$

Thus we have the theorem:

Theorem: A universe has no leakage of fields or particles into a higher dimensional space if the universe is exactly flat.

This theorem is particularly interesting in the case of our universe. It appears to be flat (or very close to flat). The flatness of our universe may be the reason no leakage of fields or particles from our universe has been detected to high accuracy.

If a universe is found with a non-zero radius of curvature then one can expect that some fields and particles may emerge from it into the Megaverse.

While a zero radius of curvature prevents the exit of fields and particles from a universe, it does not prevent the entry of mass-energy into the universe from the Megaverse. Thus our Continuous Creation Model of chapter 14 in Blaha (2017c) may be relevant. Entry is possible; exit is forbidden in this case.

Eqs. 3.1 – 3.2 are reminiscent of the 'four laws for Black Holes' which are stated later.

3.4 Quantum Fields 'Emanating' from a Universe

If the curvature of the universe is zero (open universe) then no fields emanate from it. If the curvature of the universe is non-zero (closed universe) then fields may 'leak' into the Megaverse. Then continuity conditions between a universe field and its Megaverse counterpart becomes of interest. We discuss this in detail later.

3.5 Universe Confinement by Conservation Laws

Every point in our universe is "infinitely" close to points of the Megaverse.

A universe occupies a region within the Megaverse. However because it is a lower dimension surface within the Megaverse the neighborhood of every point within a universe has an infinite number of Megaverse points that are not within the universe.

One might think that particles within a universe could then 'slip' into the Megaverse outside the universe with ease. However that is not the case. The law of momentum conservation compels particles and interactions within a universe to be confined to the universe. More importantly, the Megaverse surface tension of a flat universe confines particles and fields within a universe.[22]

The only possible ways that a particle could exit from a universe are 1) if the particle collides with a particle with a momentum, some of whose components are in Megaverse dimensions extraneous to the universe's dimensions; or 2) a particle within the universe experiences forces with components in Megaverse dimensions extraneous to the universe. We shall consider the second possibility later when we consider a mechanism for a starship to exit our universe (chapter 5). The first possibility exists if the Megaverse has a very low matter density outside of universes. The fact that this phenomenon has not been observed implies the Megaverse matter density is extremely low.

Thus conservation of momentum for particles and interactions, and surface tension, effectively confines particles within a universe even though the neighborhood of every point of a particle's trajectory contains an infinity of Megaverse points exterior to the universe. Similarly every interaction within a universe is confined when expanded in a Fourier series (assuming free fields) in universe coordinates.

It is also possible for a Megaverse particle to enter[23] a universe through perhaps a collision that results in the particle being within the universe with momentum also solely within the universe. Thus the 'point boundary' of universes is porous. Particles can enter/exit a universe under appropriate conditions.

---

[22] A useful analogy: the Megaverse is a pool of water; a universe is a denser, oil bubble within it. Surface tension caused by the cohesiveness of the oil molecules in the bubble makes it spherical (confines it to a spherical shape).

[23] In chapter 14 of Blaha (2017c) we consider a model that supplies a mechanism for the Hoyle-Narlikar continuous creation theory for the expansion of our universe. This model 'creates' mass-energy as an inflow from the Megaverse.

We conclude the 'point boundary' of a universe is not a barrier although surface tension force controls the entry/exit of particles and fields. We consider an exit mechanism from a universe in chapter5.

3.6 Objects Straddling a Universe-Megaverse Boundary

When a starship, or some other extended object, is entering/exiting a universe at some velocity the question of the state of the object arises. It is partially in and partially out of the universe. We know that the object being 4-dimensional will continue to be 4-dimensional, barring effects of forces that might "twist" parts of the object into additional dimensions.

There are also the more subtle quantum effects on the object due to the possibility of different quantization of the particles in a universe and the Megaverse. Quantization in different coordinate systems might result in different physical interpretations of matter. We resolve this issue by our PseudoQuantization Method described in Blaha (2018e). In its Appendix A we show that one can quantize using a form of Pseudoquantization that preserves (unitarily equivalent) particle interpretations in a universe and the Megaverse.

Thus extended objects can be partly in a universe, and partly in the external Megaverse without issues.

4. General Properties of the Megaverse

If one wishes to have a portrait of the Megaverse it seems likely that the it would contain universes that were scattered within it in a fashion similar to the scattering of galaxies within our universe—although on a much larger distance scale in multiple dimensions. In this chapter we will overview properties of the Megaverse and its universes. Much of this material previously appeared in Blaha (2017c) and in earlier Physics and starship travel books.

4.1 Megaverse Size, Lifetime, and Universe Separation

The first questions that naturally arise are the age, size, and general characteristics of universes within the Megaverse. In the absence of any experimental detail we will assume the relative size of entities (universes) in the Megaverse[24] is proportionate to the relative size of the entities (galaxies) in the universe.

(Average Galaxy Size)/(Universe Size) = (Average Universe Size)/(Megaverse Size)   (4.1)

Taking the average diameter of galaxies to be 400,000 light years, the age of the universe to be 13,800,000,000 light years, and the diameter of the universe to be 91.4 billion light years[25] (the estimated diameter of last scattering surface) we find the diameter of the Megaverse to be

$$\text{Diameter}_{\text{Megaverse}} = 2 \times 10^{16} \text{ light years} \qquad (4.2)$$
$$= 228{,}500 \times \text{Diameter}_{\text{Universe}}$$

---

[24] We assume the Megaverse is homogeneous at large distance scales. We also assume a proportionality between the average entity size (of galaxies in the case of unbiverses; and of umiverses in the case of the Megaverse) in part based on the relatively long lifetimes of universes and the Megaverse.

[25] There are much larger estimates of the universe's diameter based on the Inflation theory of A. Guth and others.

Using the 228,500 scale factor, we find the Megaverse age since the Megaverse purported 'Big Bang' to be

$$\text{Age}_{\text{Megaverse}} = 3 \times `10^{15} \text{ years} \tag{4.3}$$
$$= 3 \text{ million billion years}$$

If the average separation between galaxies in our universe is 3,000,000 light years, then assuming distance scaling by 228,500, the average separation between universes in the Megaverse would be

$$\text{Separation}_{\text{Universes}} = 228,500 \times 3,000,000 \text{ ly} = 7 \times 10^{11} \text{ light years} \tag{4.4}$$
$$= 700 \text{ billion light years}$$

If we now assume the mass of a universe equals the total mass-energy of our universe[26] (including Dark mass and Dark energy) which is estimated to be $m_{\text{universe}} = 3 \times 10^{54}$ kg then the gravitational potential energy between two such universes separated by 700 billion light years is

$$V = G\, m_{\text{universe}}^2 / \text{Separation}_{\text{Universes}} \tag{4.5}$$
$$= 9 \times 10^{70} \text{ kgm}^2\text{s}^{-2}$$

Then the gravitational force is

$$F = G\, m_{\text{universe}}^2 / \text{Separation}_{\text{Universes}}^2 \tag{4.6}$$
$$= 1.35 \times 10^{43} \text{ kgms}^{-2}$$

and the resulting gravitational acceleration of universes is

$$a = 4.5 \times 10^{-12} \text{ m/s}^2 \tag{4.7}$$

The Baryonic and Leptonic forces associated with the Generation group of the Unified SuperStandard Theory may slightly modify the force between the universes. We view

---

[26] We assume our universe is 'average' in mass and size for the sake of discussion.

the small acceleration between universes due to gravity as physically acceptable. In a billion years the universe velocity would be $v = 1.4 \times 0^5$ m/s with v/c = 0.0005 and the distance traveled equal to 2.2×10^{21} m = 236,540 light years – negligible compared to the Separation$_{\text{Universes}}$. Universes would, on the average, only make contact after extraordinary long times. Thus we have a preliminary coherent view of Megaverse size and distance parameters.

4.2 Likely Features of the Megaverse

There are a number of features of the Megaverse that appear to be true.

4.2.1 Megaverse Curvature

The Megaverse has gravitation. Gravitation appears to be weak in the Megaverse so it is close to a flat space. The sources of Megaverse gravitation are the mass-energy of the universes within it and the density of mass-energy of particles outside of universes. Universes have a larger relative force of gravity due to a higher mass-energy density. Universes therefore have more curvature.

4.2.2 Megaverse Time Dimension

We will assume that the Megaverse has one complex time dimension denoted y^D for the simple reason that the absence of a time dimension would make the Megaverse static.

4.2.3 Megaverse Forces

In addition to Megaverse gravitation, the Megaverse has the forces in the Unified SuperStandard Model. These forces satisfy continuity conditions at universe boundaries.

4.2.4 Megaverse Parameters

Megaverse physical constants and particle masses have the same values as in our universe due to continuity.

4.2.5 Megaverse Vacuum Fluctuations

Megaverse Vacuum fluctuations may be a source of the generation of universes and particles. Vacuum fluctuations might account for the Big Bang. The time scale for the persistence of universes generated by a vacuum fluctuation is likely to be an extrapolation of vacuum fluctuation persistence within our universe.

4.2.6 Megaverse Matter and Chemistry

The existence of many more dimensions in the Megaverse suggests that multi-dimensional forms of matter and energy could exist *between* universes. As a result such Megaverse atoms, compounds and Chemistry will be very different and much more varied than in our universe. If such matter exists in the Megaverse then 'mining' such matter for use in our universe—would give us exotic new compounds and Chemistry that would be partially inside and partially outside, of our universe.

This possibility makes venturing into the Megaverse economically and scientifically desirable since such materials cannot be created within our universe.

4.3 Features of Universes within the Megaverse

We know of our universe from the 'inside.' However the features of our universe from a Megaverse perspective are not at all certain. In this section we will describe the Megaverse view of a universe's properties.

We shall assume a universe is a closed or open surface within the Megaverse of much higher mass-energy density than the Megaverse's mass-energy density by perhaps as much as a factor of 10^{30}. Chapter 14 of Blaha (2017c) describes a model of Megaverse mass-energy inflow into our universe exemplifying this feature.

4.3.1 Universe Area and Mass

The mass of a universe is an important property since mass is one of the sources of Megaverse gravitation, and interaction between universes.

While a universe is not believed to be a black hole (although Hawking has recently jokingly? suggested that our universe may be a black hole, and even more recently suggested black holes are not quite black holes – grey?), there are general qualitative similarities that lead us to consider the possibility that the four laws of black holes[27] may apply in part (or their entirety) to universes. In particular the 2nd law states

$$dM = \kappa dA/8\pi + \Omega dJ \qquad (4.8)$$

where dM is the change in "mass/energy," A is the area of the Black Hole (universe), Ω is its angular velocity and J is the angular momentum.[28] From eq. 4.8 it appears we can reasonably define a "mass" for a universe in terms of a universe's area:

$$M = \kappa A/8\pi \qquad (4.9)$$

This definition seems to capture the physics of universes that could be used in developing a dynamics of universes as we do later. It allows us to escape the dilemma of having zero total energy for universes that would preclude treating universes as particles in the Megaverse and developing a Megaverse dynamics of universe-particles. Later we will also show how to define a mass for a universe that is time dependent.

4.3.2 Relation between Universe and Megaverse Vector Fields

In Blaha (2018e) we defined the interactions of The Unified SuperStandard Theory. These interactions and their groups also exist in the Megaverse. The Fourier expansions of fields in the Megaverse are different. They must be expressed in terms of the D coordinates of the Megaverse.

Within a universe, since each point of the universe is surrounded by Megaverse points, a field has both an expression in universe coordinates, x, and an expression in Megaverse coordinates, denoted y. For a vector field in the universe $A_U{}^\mu(x)$ there is an equivalent representation of the field in Megaverse coordinates $A_M{}^i(y)$. These representations are related by a coordinate transformation. If we define a map from universe coordinates to Megaverse coordinates with

---

[27] Wald, R. M., "The Thermodynamics of Black Holes", *Living Reviews in Relativity* **4** (6): 12119 (2001).

[28] Although the angular momentum of a universe is not measurable if there is only one universe (as DeWitt argued in a quote earlier), the existence of multiple universes within the Megaverse enables the relative angular momentum of a universe to be determined.

$$y^i = f^i(x) \tag{4.10}$$

then the field representations are related by[29]

$$A_M{}^i(y) = \partial y^i/\partial x^\mu \, A_U{}^\mu(x) \tag{4.11}$$
$$= \partial y^i/\partial x^\mu \, A_U{}^\mu(f^{-1}(y))$$

in the domain of the universe. The values of the field at Megaverse points in a neighborhood of a universe point are determined by continuity.

Outside the domain of a universe the Megaverse field value is determined by its sources.

At the boundary of a universe there must be continuity in the expectation values of the Megaverse fields.

4.3.3 Megaverse Gravitation and Free Matter

We assume that a generalization of Einstein's theory of Gravity exists in the Megaverse.

Just as our universe has matter and radiation between galaxies, it seems reasonable to assume that 'free' matter and radiation exists in the Megaverse outside of universes. Such mass-energy would have two roles: to gravitationally affect the dynamics of the Megaverse and the motion of universes within it, and to possibly fuel the expansion of universes. The expansion of our universe may be due to an influx of matter and energy from the external Megaverse. Many years ago Hoyle and Narlikar considered the possibility of 'continuous creation of matter.' We suggest that an influx of Megaverse matter may be the actual source. We considered this possibility in chapter 14 of Blaha (2017c), which contains an unpublished paper by this author written approximately seven years previously.

---

[29] Implicit in eq. 4.10 is an inverse relation $x^\mu = f^{-1}(y)$ which is necessarily based on a restriction of the y-coordinates to obtain a 1:1 relation between the y and x coordinates. The restriction is best implemented by requiring the domain of y coordinates be restricted to those y coordinates within the universe surface. The result is a 1:1 relation between the y-domain coordinates and the x universe coordinates.

Thus we arrive at a view of the Megaverse of matter and universes that is analogous to our universe of galaxies.

4.3.4 Expansion of Universes

Our universe expanded from the Big Bang to its current size and is still expanding. It is likely that other universes have undergone similar expansions. According to chapter 32 of Blaha (2018e) there is an infinite surface tension at the boundary of our universe. How can the universe have expanded, and continue expanding, under such conditions. We see a two phase expansion of the universe.

For a period of time after the Big Bang the universe did not have an infinite surface tension, thus hindering universe expansion into the Megaverse due to surface tension force. The cause is an effect discovered by Eőtvos – a temperature dependence of the surface tension force. Eőtvos pointed out a critical temperature T_c existed that caused the surface tension force to decline as the temperature increased:

$$\gamma V^{2/3} = k(T_c - T) \qquad (4.12)$$

where k is the Eőtvos constant, and V is the volume of the universe (the liquid 'drop') Assuming a spherical universe, the volume is

$$V = 4\pi R^3/3. \qquad (4.13)$$

Thus

$$\gamma = (4\pi/3)^{-2/3} k(T_c - T)R^{-2} \qquad (4.14)$$

For very high temperatures such as existed after the Big Bang $T > T_c$ γ would be negative indicating that an outward pressure existed from the universe into the Megaverse promoting expansion. Thus in the high temperature period ($T > T_c$) after the Big Bang the surface tension force favored expansion of the universe.

After this phase, the surface tension γ is positive. The Megaverse is then superficially partially 'impeding' expansion due to surface tension. However, the surface tension pressure of the Megaverse causes leakage *into* the universe from the

Megaverse causing its mass to increase, and its radius and volume to increase. Expansion due to the accretion of Megaverse mass-energy!

The above scenario is supported by the two phase model suggested by section 14.15 consisting of a Big Bang expansion model (chapters 11 – 13), and a mass-energy accretion model (chapter 14) – all of Blaha (2017c). Note as the radius of curvature goes to zero, eq. 4.14 suggests an increasing surface tension pressure 'pushing' particles into the universe.

Thus a complete universe expansion scenario is evident based on surface tension physics and the Big Bang.

4.3.5 Universe Generation from Vacuum Fluctuations

Vacuum fluctuations could generate universe-antiuniverse pairs. Antiuniverses would have certain 'negative quantum numbers.' We view universes/antiuniverses, when created (Big Bangs), as 'ultra-small' particles of large mass-energy that can proceed to expand to great size. If they are generated by a vacuum fluctuation they will, after possibly a certain time, recombine into the vacuum.

4.3.6 Universes as Black Holes

It is conceivable that a universe could be so dense and confined that it would be effectively a Black Hole. In this case the internal quantum numbers of the Black Hole universe would be inaccessible and only it's mass, velocity and angular momentum would be observables.

4.3.7 Life in Other Universes

It is likely that life exists in some if not all of the universes of the Megaverse. If the physical constants, laws, and masses of the interior of a universe are similar if not identical to ours then the possibility of intelligent species, even human-like species, is very likely. This possibility is an important motivation for humanity to reach for the Megaverse.

5. On Exiting From Our Universe

The first question on realizing that one is in a confined space is "How do I get out?" This question takes on a dramatic new aspect when we ask "How do we exit from our universe?" Like tadpoles in a pool of water it is a question of importance since universes, like pools of water, may 'evaporate' – a possibility raised from time to time by cosmologists.

In this chapter we present a mechanism for exiting our universe realizing that it could not be developed and used for perhaps *tens of thousands of years.*[30]

5.1 Looking for an Exit Point

Due to our discussion of surface tension at the boundary of the universe the first issue that we will address is whether there are 'points' where egress from our universe is facilitated. We will see that the vicinity of bodies with large gravitational fields such as neutron stars and Black Holes seriously weaken the nearby universe surface tension and are thus the best locations for exiting our universe into the Megaverse.

5.2 Neutron Star (Black Hole) Exit Point Surface Tension

Near a neutron star, or similar small, very massive body, the curvature of the universe (which is close to zero – flat) becomes significant. Then, as we saw in chapter 32, of Blaha (2018e) the surface tension force becomes possibly much smaller (not infinite!) and the force keeping mass-energy within the universe decreases enabling a starship to conveniently exit into the Megaverse.

---

[30] This time frame could be shortened if it were possible to use faster-than-light starships to execute the maneuvers described in Blaha (2018e) around a Black Hole without the destruction of the starship and its passengers. If faster than light starships become a reality then exits from the universe may become feasible only thousands of years afterwards.

In chapter 32 (Blaha (2018e)) we saw the pressure exerted by the surface tension for a 'spherical' surface area is

$$\Delta p = 2\gamma/R \qquad (5.1)$$

where R equals the radius of curvature of the surface with R = 0 for a flat space. A non-zero value for R causes the infinite force of flat space-time to become a finite force which the starship can overcome to exit.

Later in that chapter we described a slingshot maneuver for a starship in which a starship 'whips' around a neutron star using its gravity to twist into the Megaverse. The trajectory is depicted in its Fig. 5.1a. We call starships with this capability *uniships* since their ultimate goal is to travel to other universes.[31]

A neutron star (Black Hole) slingshot has an additional advantage for transit to the Megaverse. This advantage is enhanced by an ultra-fast uniship speed. An extremely large speed causes the uniship to appear 'hot' using the conversion between velocity and temperature. A high speed will create a 'hotspot' in the universe boundary surface that could substantially lower the surface tension force.[32] Eőtvos showed that the surface tension force has a critical temperature T_c which satisfies:

$$\gamma V^{2/3} = k(T_c - T) \qquad (5.2)$$

where k is the Eőtvos constant and V is the volume of the universe (the liquid 'drop'). As the uniship speed increases to an equivalent temperature beyond T_c, the Megaverse surface tension force will drop to zero, and the uniship can freely enter the Megaverse.

---

[31] The design and human factors of uniships: living conditions, travel times, communications from the Megverse, long-lived ship materials, suspended animation, and so on are described in Blaha (2017c) and earlier books.

[32] Much more importantly is the temperature in the region around a neutron star. An idea of the temperatures involved is the magnitude of the temperature in a region 1 – 7 km in size in a colliding neutron stars event. The temperature was estimated to be the extraordinary temperature 350 GK (gigakelvins) by R. Oechslin et al, arXiv:astro-ph/0507099 (2006). Temperatures of the order of 10^{12} K would seem to severely lessen or eliminate surface tension.

Figure 5.1a. The trajectory of a uniship in a slingshot maneuver with a neutron star or Black Hole (the dark circle). The repulsive baryonic force causes the "turn" away from the star/Black Hole into the Megaverse. The solid line corresponds to the time in which the uniship is wholly within the universe and dominated by gravity. The dotted line reflects the transition of the uniship into the Megaverse.

Therefore the Megaverse – universe interface can be viewed as a local surface bubbling that thins out at high speed to nothing – no barrier force.

We can estimate the critical velocity v_c as a function of the critical temperature, at which the surface tension force goes to zero, using the velocity maximum of the superluminal Maxwell-Boltzmann distribution:[33]

$$v_c \approx c + \tfrac{1}{2}\, m^2 c^5 / (2kT_c)^2 \qquad (5.3)$$

where m is the mass of the uniship (presumably hundreds of thousands of metric tons). The critical temperature can also be expected to be extremely high. As a result the ratio, the critical velocity, is likely to be large. At sufficiently high superluminal velocity there is no barrier to entry to the Megaverse.

The precise determination of the critical temperature remains to be done. Chapter 14 of Blaha (2017c) estimates the critical temperature for our models in Blaha (2018e).

---

[33] Blaha (2017c) section 1.4.3.

SECTION 2: HYBRID QUANTUM BIG_BANG-MEGAVERSE_DRIVEN THEORY

This section presents a two phase hybrid quantum theory of the Big Bang combined with a Megaverse Driven Expansion due to an influx of mass-energy from the Megaverse.

6. Universe Particle Dynamics

The origin of our universe and other universes is a weighty question that has been considered in a number of theories. Our view is that the existence of the Megaverse, and the locality of universes (in the one case of which we are certain), suggest the Megaverse is the "platform" of a type of particle physics in which universes play the role of particles. There are clear points of difference: particles have a fixed mass, particles have well-defined quantum numbers, and so on. Nevertheless we can consider universes to be variable mass particles, *universe particles*, with a dynamics that is primarily based on gravitation.

This chapter describes aspects of the interactions of universe particles due to gravitation, baryon number forces, and collisions between universes. It also the describes the genesis of universes due to vacuum fluctuations, the fission of universes, and the internal distortion of universes due to acceleration and the presence of 'nearby' universes..

6.1 The Internal Distortion of Universes

In the absence of external forces universes are considered to be uniform in the large. However the acceleration of a universe in the Megaverse can distort the universe. Also the existence of a nearby universe(s) could cause the uniformity of a universe to be lost due to gravitation and baryonic forces.[34]

6.1.1 Impact of Universe Particle Acceleration – Lopsided Internal Structure of Universe

Universes can accelerate within the Megaverse due to external Megaverse forces. Universe acceleration should be detectable within a universe as "lopsidedness."

---

[34] The baryonic forces, the Baryonic force and the Dark Baryonic force, on a universe are large due to their additivity in our universe and nearby universes.

There would be a shift of parts of the universe opposite to the direction of acceleration resulting in a difference in the features of the universe "in front" compared to those "in back" – an acceleration effect just as one sees when a jet accelerates.

Interestingly new data from the Planck observatory of the European Space Agency confirms and extends earlier data from NASA's WMAP observatory that one side of the universe appears different from the other side. There are temperature differences and mass distribution differences – just as one might expect if the universe were accelerating as a unit.

Thus we see the beginning of data suggesting our universe may be accelerating through the Megaverse. Some Planck observatory scientists have suggested their data is a preliminary indication of the Megaverse.

6.1.2 Impact of External Forces on Universe Structure

The presence of a nearby universe could cause a universe to lose its large scale uniformity and the mass-energy of the universe to drift over time to the 'nearby' side of the universe due to gravitation and baryonic forces.

6.2 Universes in Collision

We can assume that the dynamics of universes in collision will be analogous to that of galaxies in collision since gravity is a dominant force in both cases. Colliding galaxies have often been observed. Their dynamics should provide guidance for the case of universes in collision.[35]

It is clear in the case of colliding galaxies, and of colliding large nuclei (gold and lead typically) that there are several types of collisions with differing results. Similarly, the types of universe collisions can be qualitatively classified as:

1. Clean collisions in which universes nudge each other but retain their identity. These are extreme peripheral collisions. If the universes overlap slightly then the

---

[35] The high energy collision of atomic nuclei at Brookhaven, CERN and other laboratories also is analogous in overall detail with universes in collision.

typically spherical symmetry of the universes may become distorted and they may become lopsided.[36]

2. Peripheral collisions in which the universes retain their identity but are connected by a trailing string of mass-energy. Eventually the string breaks and the universes separate. Subsequently the pieces of trailing string in each universe contract due to their universe's gravitational effects and perhaps form new "bubble" universes.

3. Two universes can collide and produce multiple universes.

4. Two universes can collide in a "central" collision and amalgamate into one universe. They can intermix with both the baryonic, gauge, and gravitational forces causing a redistribution of their masses. They may separate afterwards or may coalesce into a single universe. One result of this may be lopsided universes. Our universe appears to be lopsided. Some cosmologists believe this is due to a near collision of our universe with another shortly after the Big Bang.

6.3 Creation of Universes through Gauge Field Fluctuations

One of the most exciting questions in Cosmology is the origin of our universe. The conventional view is that it originated in a Big Bang from an infinitesimal point in space. The source of the Big Bang and the prior state of the Megaverse, if there was one, is the subject of much speculation. Based on the particle interpretation of the Wheeler-DeWitt equation, the possibility of a baryonic force strongly supported by conservation of baryon number, and the Megaverse concept, it is reasonable to consider the possibility that the universe originated in a vacuum fluctuation.

In this case there would be two Big Bangs one for our universe and one for an anti-universe. One would expect that they would have opposite corresponding features:

---

[36] The Wilkinson Microwave Anisotropy Probe (WMAP) and the Planck European Space Agency satellite have been accumulating data since 2001 that suggests the universe may be lopsided with hot and cold spots on opposite sides of the universe differing from those on the other side being hotter and colder respectively—*perhaps the result of a collision when the universe was young.*

one with baryon dominance – one with anti-baryon dominance, and one left-handed – one right-handed.

Our formulation of universe particle theory allows for the generation of a universe particle and anti-particle as a vacuum fluctuation. We view a universe particle as having a substantial excess of baryons, N, as we see in our universe. Its anti-universe at the time of creation (the Big Bang point) is its "mirror image" having the "same" number of anti-baryons (baryon number –N) so that baryon number is conserved by the fluctuation event. Thus the excesses of one universe are compensated by the excesses of the other.

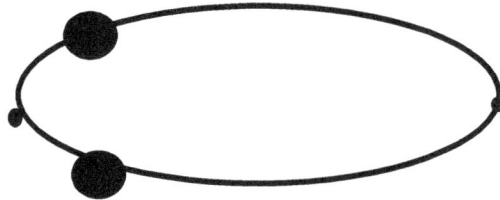

Figure 6.1. Generation of a universe – anti-universe pair as a vacuum fluctuation.

The small value of the coupling constant should lead to an extremely long lifetime for the universes generated by the fluctuation. Thus the 45 billion year life of our universe is not unreasonable. The probability of the creation of universes by vacuum fluctuations should be correspondingly small.

The sizes of the created universe and antiuniverse should be very small just as Big Bang theories of our universe suppose.

6.4 Fission of Universes

Under certain circumstances the distribution of matter in the universe may lead to the fission of the universe into two separate universes. Our theory supports this possibility for universe particles. The detailed mechanism of the fission process is not specified by the model.

6.4.1 Fission of Normal Universes

The fission of universe particles in our universe particle model is depicted in the Feynman diagram in Fig. 6.2.

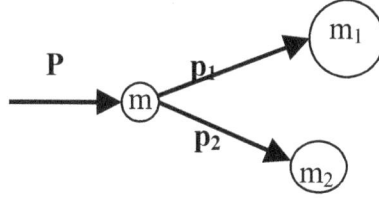

Figure 6.2. Fission of a universe particle into two universe particles.

The sum of the masses of the output universe particles is usually less than the original universe particle mass. However if the fission takes a long time and the masses are time dependent then the produced universe particles combined masses may exceed the original universe's mass.

6.4.2 Tachyon Universe Particle Fission to More Massive Universe Particles

In Blaha (2007a) we showed that a tachyonic (faster than light) particle could fission into particles of larger mass through the conversion of momentum into mass. In this section we show that a tachyonic universe particle may fission into two more massive universe particles.[37] This phenomenon is of particular interest because it enables tachyonic universes to spawn in a new novel way not previously considered in discussions of the origin of universes.

A simple model lagrangian[38] for a tachyonic universe particle is

$$\mathscr{L}_{\parallel} = \psi_T^{\ S}(Y(y))[\gamma^\mu \partial/\partial y^\mu - e_B \gamma^\mu B_{u\mu}(Y(y)) - m(t)]\psi_T(Y(y)) - \tfrac{1}{4} F_{Bu}^{\ \mu\nu}(Y(y))F_{Bu\mu\nu}(Y(y))$$
$$- \tfrac{1}{4} F_u^{\ \mu\nu}(y)F_{u\mu\nu}(y)$$

---

[37] We will use the term mass here to denote mass-energy. Since we identified mass as a multiple of area earlier, the comments here would appear to apply to universe area as well.
[38] See Blaha (2018e) and earlier books by the author for a detailed discussion.

We assume m(t) is constant.

When a particle or a universe particle fissions (decays) one normally expects that the masses of the particles or universe particles produced by the decay to be smaller than the mass of the original particle or nucleus. In the case of tachyonic (faster-than-light) elementary particles, or universe particles, a much different possibility is present: a tachyon universe can decay into heavier tachyons (perhaps through a distortion of the universe internally into two 'lumps'.) We consider the specific case of a tachyon universe particle decaying into two universe particles whose total mass is greater than the original. (See Fig. 6.3.)

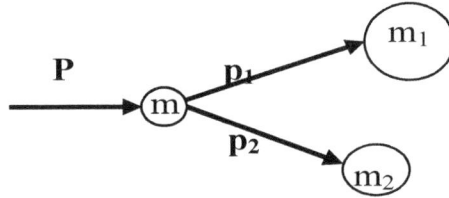

Figure 6.3. Two universe particle decay of a tachyon universe particle.

We will assume the initial tachyon universe particle has zero energy ($p^D = 0$) and thus the tachyons universe particles emerging from the decay also have total universe particle energy zero. The analysis is based on conservation of total universe energy and momentum in Megaverse space outside of universes. The below discussion applies to D-dimensional space with (D – 1)-dimensional spatial coordinates.

Momentum conservation implies

$$P = p_1 + p_2$$

Since all energies are zero

$$(cP)^2 = (c\mathbf{P})^2 = m^2$$
$$(cp_1)^2 = (c\mathbf{p_1})^2 = m_1^2$$
$$(cp_2)^2 = (c\mathbf{p_2})^2 = m_2^2$$

where $P = |\mathbf{P}|$, $p_1 = |\mathbf{p_1}|$, and $p_2 = |\mathbf{p_2}|$. If we now square the above equation for $\mathbf{P}$ and then use the above three equations we obtain

$$m^2 = m_1{}^2 + m_2{}^2 + 2m_1m_2 \cos \theta$$

where θ is the opening angle between the emerging universe particles momenta of $\mathbf{p_1}$ and $\mathbf{p_2}$. There are a number of interesting cases:

Case $\theta = 0$:

$$m = m_1 + m_2$$

The masses of the outgoing universe particles sum to the mass of the original tachyon universe particle.

Case $\theta = \pi/2$:

$$m^2 = m_1{}^2 + m_2{}^2$$

The masses of each outgoing universe particle tachyon are less than the mass of the original tachyon universe particle.

Case $\theta = \pi$:

$$m^2 = (m_1 - m_2)^2$$

In this case either $m_1 > m$ or $m_2 > m$. Thus one of the outgoing tachyon universe particles has a greater mass than the original tachyon universe particle. Mass is effectively created from the spatial momentum of the initial universe particle. This process is the inverse of normal particle and universe particle fission where the sum of the outgoing masses is always less than the original particle's mass and the difference is mass converted into energy in the form of additional photons.

This last case, where one of the outgoing universe particles is more massive than the original universe particle, is not just for $\theta = \pi$. Since

$$\cos \theta = (m^2 - m_1^2 - m_2^2)/(2m_1m_2)$$

we see that the sum of the outgoing universe particle masses is always greater than the original tachyon universe particle *mass (except when $\theta = 0$)* since

$$\cos \theta = 1 + [m^2 - (m_1 + m_2)^2]/(2m_1m_2) \leq 1$$

and thus

$$[m^2 - (m_1 + m_2)^2]/(2m_1m_2) \leq 0$$

Note $m = m_1 + m_2$ only if $\theta = 0$.

Since we can transform the above discussion to the case of universe particle tachyons having non-zero Megaverse energy using an ordinary D-dimensional Lorentz transformation, the discussion in this subsection is general.

We therefore conclude that when a tachyon universe particle decays into two tachyon universe particles the sum of the masses of the produced tachyon universe particles is greater than the mass of the original tachyon universe particle except if the angle between the momenta of the produced tachyon universe particles is zero. In that case the sum of the masses of the produced tachyon equals the mass of the original tachyon universe particle and the produced universe particles overlap.

7. A Driving Force for Universe Expansion

Some years ago the author introduced a new method[39] that ensured perturbation theory calculations were always finite to all orders in perturbation theory. The method was based on a generalization of space-time coordinates to have imaginary q-number parts that consisted of a massless vector quantum field theory that was similar in form to the quantum electromagnetic field. We called these coordinates Two-Tier coordinates because they were based on an extended two level extremum principle.

We showed that these quantum smeared coordinates of Two-Tier Quantum Field Theory succeeded in removing all ultra-violet infinities in perturbation theory including the fermion triangle infinities. Remarkably the high precision, low energy[40] predictions of QED remained true in Two-Tier QED and thus remained consistent with experiment to a hitherto unsurpassed level of accuracy. 'Low' energy predictions in other quantum field theories also remained unchanged. At high energies, Two-Tier perturbation theory results are finite and consequently all ultra-violet infinities, to any order in perturbation theory, in *any number of space-time dimensions* were eliminated.

7.1 Additional Benefits of Two-Tier Coordinates

In addition to removing perturbation theory infinities Two-Tier coordinates have the features:

1. They enable us to define finite theories of Quantum Gravity and 'non-renormalizable' quantum field theories based on polynomial lagrangians.

2. They enable us to tame vacuum fluctuations.

---

[39] See Blaha (2002) and (2005a).
[40] Relative to a mass scale that was perhaps of the order of the Planck mass.

3.   They enable us to eliminate infinities associated with the Big Bang.

4.   They provide a mechanism to generate the explosive growth of the universe in a role as part of Dark Energy.[41]

Items 3 and 4 were proved in earlier books.[42] They will be part of a new extended theory of universes in this book. We will see Two-Tier Quantum Field Theory is required at the most fundamental level.

7.2 Two-Tier Features in 4-Dimensional Space-Time

Two-Tier Quantum Field Theory,[43] is based on a new method in the Calculus of Variations. It uses two 'layers' of fields to introduce quantum coordinates. We shall consider this technique briefly using a massless vector field $Y^\mu(y)$ analogous to the electromagnetic field.

In 4-dimensional space-time the massless vector field has the form $Y^\mu(y)$ where the index μ ranges from 0 through 3. The X^μ coordinate system, where it appears, has a c-number real part and a q-number imaginary part. Thus particle fields which are normally defined on four-dimensional real space-time are now defined on a complex four-dimensional space-time where four imaginary dimensions will appear as *Quantum Dimensions* embodied in a vector quantum field $Y^\mu(y)$:

$$X^\mu(y) = y^\mu + i\,Y^\mu(y)/M_c^2 \qquad (7.1)$$

where M_c is an extremely large mass of the order of the Planck mass or perhaps much larger.

---

[41] See Blaha (2017b) and earlier books for details. This section is basically a summary of some features.
[42] See Blaha (2004): *Quantum Big Bang Cosmology.*
[43] See Blaha (2005a), and Blaha (2002), for discussions of this new method to eliminate infinities in quantum field theory calculations.

The $Y''(y)$ field is a function of the subspace y coordinates. The real part of the space-time dimensions will be taken to be the space of real-valued y coordinates.[44]

The imaginary part of space-time coordinates is the massless $Y''(y)$ vector quantum field that is suppressed further by a very large mass scale that reduces the imaginary Quantum Dimensions to the infinitesimal except at large momenta. The effects of Quantum Dimensions only become appreciable in quantum field theory at energies of the order of M_c. At these energies exponential Gaussian factors in each particle (and ghost) propagator are generated by the Quantum Dimensions and serve to make perturbation theory calculations ultra-violet finite – including calculations in Quantum Gravity.

The formalism introduces a new form of interaction that does not have the form of the simple polynomial interactions that have hitherto dominated quantum field theories. This form of interaction takes place via the composition of quantum fields and can be called a *Dimensional Interaction* or an *Interdimensional Interaction* since it affects particle behavior through Quantum Dimensions.

The basic ansatz of the Two-Tier formalism is to replace every appearance of a coordinate x in a quantum field with the variable

$$x^\mu \rightarrow X^\mu = (y^0, \mathbf{y} + \mathbf{Y}(y^0, \mathbf{y})/M_c^2) \tag{7.2}$$

where $\mathbf{Y}(y^0, \mathbf{y})$ is the spatial part of a free massless vector field with features that are identical to the free QED field in the Radiation gauge.

Then one finds that the momentum space free field Feynman propagators G(k) of all particles acquires a Gaussian factor $\exp(h(k))$:

$$G(k) \rightarrow G(k)\exp(h(k)) \tag{7.3}$$

so that all perturbation theory diagrams are finite. The result is finite perturbative results for all calculations to any order in perturbation theory.

---

[44] In a deeper theory the real part might also be a quantum field that undergoes a condensation to generate c-number coordinates. We will not consider this possibility in this book.

Blaha (2005a) shows that Two-Tier theories are finite, Poincare covariant, and unitary. (See Blaha (2005a), chapter 5, for a complete discussion.)

7.3 Importance of Two-Tier Coordinates

Two-Tier coordinates can be used in an extension of the Robertson-Walker Model to create a Big_Bang sector that prevents infinities at the Big Bang point in time. The fuzziness of these coordinates effectively prevents a "condensation" of a universe to a point going backward in time from the present as many current Big Bang models do. See the following chapters for details.

8. Plan for the Hybrid Big_Bag-Steady_State Theory Presentation

The plan of the development of the theory begins with the Robertson-Walker Model in real space-time. We then proceed to generalize the Robertson-Walker Model to have q-number coordinates consisting of a real-valued part and an imaginary part that is a vector field (similar to free QED) that introduces a repulsive effect at extremely short distances. This effect prevents a collapse at the Big_Bang to a point with infinite mass-energy and temperature.

The motivation for this construct is analogous to the motivation that appeared when atomic models were being constructed at the beginning of Quantum Mechanics. Quantum Mechanics prevented the collapse of an atom to a point as circling classical electrons spiraled to the central nucleus. In the present case the universe is prevented from beginning from a point by the imaginary quantum field that provides a form of "anti-gravity" at ultra-short distances.

The introduction of an imaginary quantum part of coordinates is not a form of "Dark Energy" and yet it strongly influences the initial start and growth of the universe.

Having defined a form of quantum coordinates for the Robertson-Walker Model, we then proceed to calculate the growth of the universe from t = 0 using the known mass-energy with the proviso that we treat Dark Matter as similar to ordinary matter in its effect on universe growth (the Hubble constant).

We also introduce a new term in the total mass-energy for the effect of universe surface tension at the boundary of the Megaverse. In doing this, we note that every point of the universe is in a Megaverse neighborhood. Thus the surface tension energy contribution pervades the entire universe and causes an inflow of Megaverse mass-energy throughout the universe. As a result the universe expands more rapidly and the Hubble constant increases with time.

We calculate the Hubble Constant for all time starting as finite at t = 0 and growing in time as we presently observe. We also calculate the radius and temperature of the universe in the large as a function of time. Much of this theory appeared in Blaha (2004).

9. Mass-Energy Growth Factors for Universes

In chapter 6 we discussed some possible seeds from which universes might spring. One important possibility is as a vacuum fluctuation that produces a universe-antiuniverse pair. An attractive feature of this mechanism is that it accounts for the preponderance of masses in our universe as well as other features such as left-right asymmetry. There are other proposals for the origin of our universe that merit attention such as an oscillating universe scenario.

Whatever the source of our universe, it is clear that it began as a very small region (or a point) containing a large amount of mass-energy. So we begin with that picture. We then develop a theory of how the universe evolved to its current enormous (and growing) size. The theory will be based on a generalized Robertson-Walker Model described beginning in the next chapter that incorporates complex valued coordinates.[45]

The propellants for the expansion of the universe are: 1) its mass-energy contents; 2) Quantum effects of Two-Tier coordinates; and 3) an influx of mass-energy into the universe from the Megaverse. Together these factors combine to give an initial universe metastable state of finite size, to generate growth as represented by the Hubble constant, and to cause the Hubble constant to increase with time (as appears experimentally). If any of these three growth elements are absent, it becomes very difficult to understand the observed growing size of the Hubble constant. Currently there are numerous speculations as to the origin of the Hubble constant growth. Most of these speculations are complex and somewhat hypothetical. This quantum model leads clearly to the observed universe and its growth.

Our relatively simple model resolves the mystery of Hubble constant growth.

---

[45] We describe aspects of Complex General Relativity in Blaha (2018e).

9.1 Quantum Effects of Complex Two-Tier Coordinates

We believe that our universe, other universes, began with a Big Bang at some point in the distance past. The universe initially was extremely small and perhaps could have been a mathematical point. In any case it had an extremely large mass-energy that powered a great expansion. Chapter 7 discussed the importance of Two-Tier quantum coordinates in affecting the growth of the universe—especially in its initial stages at t = 0, although they have an effect up to the present. Thus Two-Tier quantum coordinates are a significant part of our model although they do not directly contribute to the mass-energy fueling universe expansion.

We will now describe the components of a universe's mass-energy driving expansion. We assume the universe is uniform in the large throughout its lifetime.

9.2 Universe Energy Density Components

The universe, as we know it today, contains a variety of forms of energy. The current densities of these forms of energy will be seen later. From them we can develop the form of the total energy density and project it back to the instant of the Big Bang. The Big Bang period is significantly different as we will see. The total energy density is composed of three major parts: energy density, mass density and Megaverse mass-energy inflow, which all depend on time t.

9.2.1 "Electromagnetic" Energy Density

The universe's "Electromagnetic" energy density $\Omega_\Gamma = \rho_\Gamma / \rho_{crit}$ is composed of two parts; the "visible" energy density Ω_γ and the "Dark" energy density Ω_Λ, which makes its presence known indirectly in astrophysical observations.[46]

$$\Omega_\Gamma(t) = \Omega_\gamma(t) + \Omega_\Lambda(t) \tag{9.1}$$

where ρ_Γ is the universe's electromagnetic energy density, and ρ_{crit} is the critical density of the universe.

---

[46] ρ_{crit} is the critical density of the universe. $\rho_{crit} = 1.05371(5) \times 10^{-5}$ h^2 (GeV/c²) cm⁻³ where h =0.678(9). From M. Tanabashi *et al* (Particle Data Group), Phys. Rev. D**98**, 030001 (2018).

9.2.2 Mass Density

The universe's mass density $\Omega_M = \rho_M/\rho_{crit}$ is also composed of two parts; the "visible" mass density Ω_m and the "Dark" mass density Ω_d, which makes its presence known indirectly in astrophysical observations. In specifying Ω_M we assume the Dark and normal mass densities have the same time dependence and gravitational characteristics unlike some other models of universe expansion:

$$\Omega_M(t) = \Omega_m(t) + \Omega_d(t) \tag{9.2}$$

9.2.3 Inflowing Surface Tension "Energy" Density Ω_T

The third major contribution to the total mass-energy density is an inflow of mass-energy from the Megaverse to every point in the universe. As we described in Blaha (2018e), and earlier books, each point in the universe is within a neighborhood of Megaverse points. Thus each point in the universe is on the universe's boundary with the Megaverse. There is an inherent inward directed surface tension at each point due to the gravitation of the mass-energy within the universe. The gravitational force is directed towards the "center" of the universe just as the earth's gravity is cumulatively directed towards the center of the earth.

The surface tension force causes Megaverse mass-energy to enter the universe in extremely minute amounts per unit volume. However the enormity of the universe creates a great continuing increase in the universe's mass-energy.[47]

The ingredients of the surface tension energy density inflow are:

1. The gravitational (surface tension) force on the universe's surface is

$$\kappa = \rho_M(t)V(t) = 2\pi^2 G\rho_{tot}(t)R(t) \tag{9.3}$$

---

[47] Hoyle and Narlikar proposed a Steady State model of the universe of this sort. However they could not account for the source of the influx since they were unaware of the Megaverse and the consequent surface tension effect. See H. Bondi and T. Gold, M.N.R.A.S. **108**, 252 (1948); F. Hoyle, M.N.R.A.S. **108**, 372 (1948); F. Hoyle and J. V. Narlikar, Proc. R. Soc. London **A290**, 162 (1966) and references therein.

in the Robertson-Walker model where $\rho_M(t)$ is the universe average density.

2. The surface tension γ of the universe is the Eőtvős-like expression:[48]

$$\gamma = k_E T_H/(4\pi R^2)^{-1} \qquad (9.4)$$

where

$$T_H = \kappa/2\pi$$

where T_H is the Hawking black hole temperature, k_E is a constant, and R is the radius of curvature. In the Robertson-Walker model the radius is $R = k^{-\frac{1}{2}}a(t)$ where k is the Robertson-Walker constant.

Putting these concepts together we estimate the surface tension mass-energy density $\Omega_T = \rho_T/\rho_{crit}$ inflow per unit time at time t to be

$$\Omega_T(t) = C_T G(\rho_{tot}(t)/R)\,\rho_{Mega} \qquad (9.5)$$

where C_T is a constant, ρ_{tot} is the current total mass-energy density of the universe, and ρ_{Mega} is the total mass-energy density of the Megaverse and assumed to be constant.

9.3 Total Universe Mass-Energy Density

Taking account of the aforementioned features the total energy density of the universe at a given moment is

$$\rho_{tot}(t) \equiv \rho_{crit}\Omega_{tot}(t) = \rho_{crit}[\Omega_\Gamma(t) + \Omega_M(t) + \Omega_T(t)]$$

$$= \rho_{crit}[\Omega_\gamma(t) + \Omega_\Lambda(t) + \Omega_m(t) + \Omega_d(t)] + C_T G\rho_{Mega}\int_{t_T}^{t} dt'\rho_{tot}(t')/R(t')$$

$$(9.6)$$

assuming that the Megaverse mass-energy influx began at a time $t = t_T$.

---

[48] Callaway, D., Phys. Rev. E **53**, 3738 (1996).

In chapter 12 we will exhibit the time dependence of the quantities in eq. 9.6. and calculate the Hubble constant in chapter 19. Eq. 9.6 can be expressed as closed form solutions. We note

$$\rho_{tot}(t) = \rho_{crit}[\Omega_\gamma(t) + \Omega_\Lambda(t) + \Omega_m(t) + \Omega_d(t) + \Omega_T(t)] \equiv \rho_{crit}[\Omega_H(t) + \Omega_T(t)] \quad (9.7)$$

We define

$$C_M = C_T G \rho_{Mega} \quad (9.8)$$

and then take the derivative of eq. 9.6 to obtain

$$d\rho_{tot}(t)/dt = \rho_{crit}d\Omega_H(t)/dt + C_M \rho_{tot}(t)/R(t) \quad (9.9)$$

with

$$\Omega_H(t) = \Omega_\gamma(t) + \Omega_\Lambda(t) + \Omega_m(t) + \Omega_d(t) \quad (9.10)$$

Solving eq. 9.9 we obtain

$$\rho_{tot}(t) = \rho_{crit}\exp\left[C_M\int_{t_T}^{t}dt'/R(t')\right] \int_{t_T}^{t}dt'' \exp\left[-C_M\int_{t_T}^{t''}dt'/R(t')\right] d\Omega_H(t'')/dt'' \quad (9.11)$$

In chapter 14 we will calculate the total energy-momentum density as a function of time to obtain expressions for the scale parameter of the Robertson-Walker model and the resulting Hubble Constant.

10. Derivation of an Extended Robertson-Walker Model

This chapter provides the derivation of a generalization of the Roberson-Walker solution of General Relativity for use in our discussion of the Quantum Big Bang Theory originally presented in Blaha (2004). This chapter is extracted from chapter 8 of Blaha (2004).

The generalization derived here has not been derived before, to our knowledge, perhaps because, at the level of c-number General Relativity, it is fully equivalent to the known Robertson-Walker model. However when the theory has q-number parts introduced in a physically meaningful way, as we do, it is no longer equivalent to the c-number Robertson Walker model. The essential benefit of this quantized Robertson-Walker Model is that it supports a scale factor that is a function of both time and space. The dependence on space (radial distance) enables the model to avoid infinities at $t = 0$.

10.1 The Robertson-Walker Metric

Much of the current modeling of the evolution and properties of the universe is based on the assumption of a Robertson-Walker metric which is used in the Einstein equations to obtain a first order differential equation for the scale factor $R(t)$:

$$\dot{R}^2 + k = 8\pi G \rho R^2 / 3 \qquad (10.1.1)$$

where k is a factor in the three-dimensional spatial curvature of the Robertson-Walker metric. The radius of curvature is

$$K_3(t) = k / R^2(t) \qquad (10.1.2a)$$

and the volume of the universe is

$$V = 2\pi^2 R^3 \qquad (10.1.2b)$$

Eq. 10.1.2 suggests that accurate measurements of the Hubble constant and other cosmological quantities could lead to an accurate determination of the curvature, the time dependence of the Hubble constant, and of R(t).

The form of the Robertson-Walker metric (We consider a real space-time only in this chapter) follows from the assumption of a maximally symmetric three-dimensional subspace whose metric has eigenvalues of the same sign (negative in our formalism) residing within a four-dimensional space-time with one positive eigenvalue and three negative eigenvalues. A maximally symmetric space is isotropic and homogeneous.[49]

Although the Robertson-Walker metric does not appear to embody the concept of an absolute space-time the general arguments presented in chapter 2 of Blaha (2004) indicate that it does in fact implicitly define an absolute reference frame.[50]

Therefore it is sensible to inquire whether a more general metric – a generalization of the Robertson-Walker metric – might be worth investigating – particularly in view of the major unexplained mysteries of Dark Energy as well as other new data showing the existence of massive black holes and quite mature galaxies shortly after (two or three billion years) the origin of the universe. The pile-up of mysteries from WMAP, SDSS and other sources indicates a reconsideration of fundamental assumptions may be worthwhile.

Therefore we will examine the "simplest" generalization of the Robertson-Walker metric in this chapter with a view towards elucidating some of these mysteries. More importantly, we will use this generalization in chapter 19 and subsequent chapters to develop a non-singular, Two-Tier formulation of the dynamics of the universe "at the beginning of time" – the Big Bang – taking account of quantum effects.

From the point of view of the definition of maximally symmetric subspaces the most immediate generalization of the Robertson-Walker metric is to assume a

---

[49] See Chapter 13 of Weinberg (1972) for a detailed discussion.

[50] This conclusion is now obvious due to the existence of the Megaverse. We note there can only be one absolute reference frame up to a Lorentz transformation since the sets of inertial reference frames of two absolute reference frames must be the same. This implies any two absolute reference frames must be related by a Lorentz transformation since absolute frames are necessarily flat inertial frames.

maximally symmetric *two-dimensional* subspace within a four-dimensional space-time. The general form of the metric in this case is:

$$d\tau^2 = A_{tt}(t, r) \, dt^2 + 2A_{rt}(r,t) \, dt \, dr + A_{rr}(t, r) \, dr^2 + B(r,t)(d\theta^2 + \sin^2\theta d\varphi^2) \quad (10.1.3)$$

where A_{ik} is a 2×2 symmetric matrix with one positive eigenvalue and one negative eigenvalue, and $B(r, t)$ is a negative function of r and t.

10.2 A Generalization of the Roberson-Walker Metric

We shall consider a generalization of the Robertson-Walker metric (eq. 5.5.1 of Blaha (2004)), which is a special case of eq. 10.1.3 that preserves the overall form of the Robertson-Walker metric but allows the scale factor a(t) to depend on r as well as t:

$$R(t) \rightarrow A_0(t, r) \qquad (10.2.1)$$

The generalized metric that we will analyze is embodied in the invariant interval expression:

$$d\tau^2 = dt^2 - A_0^2(t, r)[dr^2/(1 - kr^2) + r^2(d\theta^2 + \sin^2\theta \, d\varphi^2)] \qquad (10.2.2)$$

The introduction of a dependence on the radius r in $A_0(t, r)$ in eq. 10.2.2 eliminates the homogeneity of the three-dimensional spatial subspace reducing it to a two-dimensional maximally symmetric subspace.

We note that the general solution of the Einstein equations for the standard case of a perfect fluid lead to the usual view of the expansion of the universe (after the Big Bang Epoch), Hubble's law, the red shifts of radiation from distant sources of radiation, and the Cosmic Microwave Background (CMB) radiation.

10.3 The Einstein Equations for the Generalized Robertson-Walker Metric

The Einstein equations can be written

$$R_{\mu\nu} = -8\pi G S_{\mu\nu} \qquad (10.3.1a)$$

$$S_{\mu\nu} = T_{\mu\nu} - \tfrac{1}{2} g_{\mu\nu} T^{\sigma}{}_{\sigma} \qquad (10.3.1b)$$

where $R_{\mu\nu}$ is the Ricci tensor and $T_{\mu\nu}$ is the energy-momentum tensor. Assuming the energy-momentum tensor has the form of the energy-momentum tensor of a perfect fluid with the only non-zero components:

$$T_{tt} = \rho \, g_{tt} \qquad (10.3.2)$$

$$T_{rr} = -p g_{rr} \qquad (10.3.3)$$

$$T_{\theta\theta} = -p g_{\theta\theta} \qquad (10.3.4)$$

$$T_{\phi\phi} = -p g_{\phi\phi} \qquad (10.3.5)$$

where ρ is the density and p is the pressure, and the non-zero components of $S_{\mu\nu}$ are:

$$S_{tt} = \tfrac{1}{2} (\rho + 3p) \, g_{tt} \qquad (10.3.6)$$

$$S_{rr} = -\tfrac{1}{2} (\rho - p) g_{rr} \qquad (10.3.7)$$

$$S_{\theta\theta} = -\tfrac{1}{2} (\rho - p) g_{\theta\theta} \qquad (10.3.8)$$

$$S_{\psi\psi} = -\tfrac{1}{2} (\rho - p) g_{\phi\phi} \qquad (10.3.9)$$

In particular, the fact that

$$S_{tr} = 0 \qquad (10.3.10)$$

for a perfect fluid results in an important simplification in the solution of the Einstein equations for this case.

The density $\rho = \rho(t)$ and the pressure $p = p(t)$ are assumed to be solely functions of time t as is usual in the case of a perfect fluid.

10.4 Differential Equations for the Generalized Scale Factor A(t, r)

The dependence of the scale factor A_0 on both r and t leads to a significantly more complicated calculation of the Ricci tensor. We start by noting

$$g_{tt} = 1 \qquad g_{rr} = -A_0^2/(1 - kr^2) \qquad g_{\theta\theta} = -A_0^2 r^2 \qquad g_{\phi\phi} = -A_0^2 r^2 \sin^2 \theta \qquad (10.4.1)$$

Despite the dependence of A_0 on both t and r in the generalized case a direct calculation of the tt Ricci tensor component yields the familiar expression:

$$R_{tt} = g_{tt} \, 3 \ddot{A}_0 / A_0 \qquad (10.4.2)$$

where we use dots over A_0 to indicate partial derivatives with respect to time:

$$\ddot{A}_0 \equiv \partial^2 A_0 / \partial t^2 \qquad (10.4.3)$$

However, the tr-component of the Ricci tensor R_{tr}, which is zero in the case of the ordinary Robertson-Walker metric, is non-zero in the more general case under consideration:

$$R_{tr} = 2 \, \partial(\partial A_0/\partial t)/\partial r \qquad (10.4.4)$$

The corresponding Einstein equation is

$$R_{tr} = 2 \, \partial(A_0^{-1} \, \partial A_0/\partial t)/\partial r = -8\pi G S_{tr} = 0 \qquad (10.4.5)$$

for a perfect fluid. Eq. 10.4.5 implies that $A_0(t, r)$ factorizes:

$$A_0(t, r) = a(t) b_0(r) \qquad (10.4.6)$$

This factorization results in a substantial simplification in the non-linear Einstein equations considered next, which are shown to be separable in the radial and time variables.

Before proceeding to the consideration of the remaining Einstein equations it is convenient to redefine the radial coordinate using

$$\check{r} \, b(\check{r}) = r \, b_0(r) \qquad (10.4.7)$$

and

$$\text{dř/dr} = b_0(r)[b(ř)(1 - kr^2)^{1/2}]^{-1} = ř\,[r(1 - kr^2)^{1/2}]^{-1} \qquad (10.4.8)$$

While this change of coordinates does not change the physical content of the theory it does lead to simpler Einstein equations. The change of radial coordinate results in a new form of the invariant interval:

$$d\tau^2 = dt^2 - a^2(t)b^2(ř)[dř^2 + ř^2(d\theta^2 + \sin^2\theta\,d\varphi^2)] \qquad (10.4.9)$$

The new radial coordinate ř is related to the old radial coordinate by

$$ř = \{[1 - (1 - kr^2)^{1/2}]/[1 + (1 - kr^2)^{1/2}]\}^{1/2} \qquad (10.4.10)$$

and

$$r = 2k^{-1/2}ř(1 + ř^2)^{-1} \qquad (10.4.11)$$

Note the range of ř is [0, 1].

A direct calculation of the Ricci tensor for the metric in eq. 10.4.9 with $A(t, ř)$ defined as:

$$A \equiv A(t, ř) = a(t)b(ř) \qquad (10.4.12)$$

leads to the following Einstein equations (remembering our flat space cartesian metric is $\eta_{tt} = +1$ and $\eta_{ij} = -\delta_{ij}$ for i, j = spatial indices):

$$R_{tt} = g_{tt}3\ddot{A}/A = -8\pi GS_{tt} = -4\pi G(\rho + 3p) \qquad (10.4.13)$$

$$R_{tř} = 2\,\partial(A^{-1}\,\partial A/\partial t)/\partial ř = -8\pi G\,S_{tř} = 0 \qquad (10.4.14)$$

$$R_{řř} = g_{řř}[\ddot{A}/A + 2(\dot{A}/A)^2 - 2(řA^2)^{-1}\,\partial(ř\,A'/A)/\partial ř]$$
$$= -8\pi G\,S_{řř} = 4\pi G\,g_{řř}(\rho - p) \qquad (10.4.15)$$

$$R_{\theta\theta} = g_{\theta\theta}[\ddot{A}/A + 2(\dot{A}/A)^2 - A''/A^3 - 3A'/(řA^3)^{-1}]$$

$$= -8\pi G \, S_{\theta\theta} = 4\pi G \, g_{\theta\theta}(\rho - p) \tag{10.4.16}$$

$$R_{\phi\phi} = g_{\phi\phi}[\ddot{A}/A + 2(\dot{A}/A)^2 - A''/A^3 - 3A'/(\check{r}A^3)^{-1}]$$

$$= -8\pi G \, S_{\phi\phi} = 4\pi G \, g_{\phi\phi}(\rho - p) \tag{10.4.17}$$

where

$$A' = \partial A/\partial\check{r} \tag{10.4.18}$$

and

$$A'' = \partial^2 A/\partial\check{r}^2 \tag{10.4.19}$$

10.5 The Solution for the Generalized Scale Factor A(t, r)

We begin by substituting eq. 10.4.13 in eq. 10.4.15, and then substituting the factorization of A (eq. 10.4.12). The result is a separable equation:

$$\dot{a}^2 - 8\pi G\rho a^2/3 - (\check{r}b^2)^{-1} \, \partial(\check{r} \, b'/b)/\partial\check{r} = 0 \tag{10.5.1}$$

Since the first two terms in eq. 10.5.1 are solely functions of t while the third term is solely a function of ř we obtain the separated equations:

$$\dot{a}^2(t) - 8\pi G\rho a^2(t)/3 = -k \tag{10.5.2}$$

where $\rho = \rho_{tot}$ and

$$k + (\check{r}b^2)^{-1} \, \partial(\check{r} \, b'/b)/\partial\check{r} = 0 \tag{10.5.3}$$

where k is a separation constant, which we provisionally identify as the curvature parameter k in eq. 10.2.2.

Eq. 10.5.2 is precisely the equation used for the time dependent scale factor in current cosmological models (eq. 5A.3.2) using the Robertson-Walker metric. Therefore its solution, which depends on the time dependence of the energy density, will be the same as that of the corresponding conventional cosmological model for the same density.

Eq. 10.5.3 is a differential equation for the *spatial* expansion scale factor $b(\check{r})$. Under the assumption of a perfect fluid it depends solely on the constant k and is independent of the details of the perfect fluid (i.e. its density and pressure). There are two solutions of the second order non-linear differential equation for $b(\check{r})$. These solutions can be written as

$$b_1(\check{r}) = 2\gamma\delta k^{-\frac{1}{2}}[\delta^2 + \gamma^2 \check{r}^2]^{-1} \tag{10.5.4}$$

where γ and δ are constants, and

$$b_2(\check{r}) = \sigma k^{-\frac{1}{2}}[\check{r}\,(\varsigma \pm i\sigma \ln \check{r})]^{-1} \tag{10.5.5}$$

where σ and ς are constants.

$b_1(\check{r})$ (in eq. 10.5.4) is the only physically acceptable solution for the case of a perfect fluid with energy-momentum tensor specified by eqns. 10.3.2 – 10.3.5. Reason: The solution $b_1(\check{r})$ satisfies eqns. 10.4.16 and 10.4.17 while the other solution $b_2(\check{r})$ does not satisfy these equations. A necessary and sufficient condition for $b(\check{r})$ to satisfy eqns. 10.4.16 and 10.4.17 is that

$$b'' - \check{r}^{-1} b' - 2 b'^2/b = 0 \tag{10.5.6}$$

where ´ denotes a derivative with respect to $\check{r}$. Eq. 10.5.6 follows from subtracting the coefficients of the metric tensor component factors in eq. 10.4.15 from the coefficients of the metric tensor component factors in eq. 10.4.16 (or eq. 10.4.17). The solution $b_1(\check{r})$ satisfies eq. 10.5.6 for all values of γ and δ. The solution $b_2(\check{r})$ does not satisfy eq. 10.5.6 for any choice of a and β except the trivial choice $a = 0$ and is therefore physically irrelevant. It might have some relevance in the case of a non-perfect fluid. We will not investigate that possibility in the present work.

10.6 The Solution Expressed in the Original Radial Coordinate

The solution that we have obtained for the generalized Roberson-Walker case with radial coordinate $\check{r}$ can be related back to the generalized Robertson-Walker solution using the original radial coordinate r. Eq. 10.4.7 implies

$$b_0(r) = \check{r}\, b(\check{r})/r \qquad (10.6.1)$$
$$= 2a\,[1 + a^2 + (1 - a^2)(1 - kr^2)^{1/2}]^{-1} \qquad (10.6.2)$$

using eqns. 10.4.10 and 10.5.4, and defining the constant α as

$$\alpha = \gamma/\delta \qquad (10.6.3)$$

An important special case of eq. 10.6.2 is the case where $a = 1$ (Ockham's Razor!). In this case we find eq. 10.6.2 becomes

$$b_0(r) = 1 \qquad (10.6.4a)$$

and

$$A_0(t, r) = a(t) \qquad (10.6.4b)$$

thus recovering the normal Robertson-Walker solution exactly as a special case from eqns. 10.2.2, 10.4.6, and 10.5.2. In this case we see

$$b(\check{r}) = 2k^{-1/2}(1 + \check{r}^2)^{-1} \qquad (10.6.5)$$

from eq. 10.5.4 in the $\check{r}$, θ, ϕ coordinate system. We considered this case within the expanded framework of a quantized model of the beginning of the universe in chapter 19 where it has non-trivial consequences.

10.7 Equivalence of the General Solution with the Original Robertson-Walker Solution

The general solution of the Einstein equations in the case of a perfect fluid (eqns. 10.4.13 – 10.4.17) for the scale factor $A(t, \check{r})$ in the generalized metric specified by eq. 10.4.9

$$d\tau^2 = dt^2 - A^2(t, \check{r})[d\check{r}^2 + \check{r}^2(d\theta^2 + \sin^2\theta\, d\varphi^2)] \qquad (10.4.9)$$

is given by

$$A(t, \check{r}) = a(t)b(\check{r}) = 2ak^{-1/2}a(t)[1 + a^2\check{r}^2]^{-1} \qquad (10.7.1)$$

where a is a constant given by eq. 10.6.3 and where a(t) is the solution of the standard equation (eq. 10.5.2) for the time dependent scale factor in the case of the Robertson-Walker metric. We now note that if we define a new radial vector

$$\mathfrak{r} = a\check{r} \tag{10.7.2}$$

then

$$d\tau^2 = dt^2 - a^2(t)4k^{-1}[1 + \mathfrak{r}^2]^{-1} [d\mathfrak{r}^2 + \mathfrak{r}^2(d\theta^2 + \sin^2\theta \, d\varphi^2)] \tag{10.7.3}$$

Comparing this invariant interval expression with the $a = 1$ expression for $b(\check{r})$ given in eq. 10.6.5 we conclude that we have proved the following theorem:

Theorem: The solution of the Einstein equations for the case of a perfect fluid for the generalized Robertson-Walker metric (eq. 10.2.2) is equivalent to a solution of the Einstein equations for the case of a perfect fluid in the case of the Robertson-Walker metric with a scale factor a(t) that is solely dependent on time.

In the case of *classical* gravitation theory the solutions are fully equivalent and related by a simple change of radial coordinates. Thus the homogeneity condition that we had relinquished at the beginning of our discussion is reinstated and the solution of the generalized case is consistent with a *maximally symmetric three-dimensional space*. The origin of the coordinate system can be chosen to be any point in space.

In the case of quantized versions of the Robertson-Walker model and our generalization of it we will see that the solutions are *generally not equivalent*. We explored a particular example of a quantized gravitational model that illustrates this point in chapter 8. The quantized model, which we defined, should be viewed as a first attempt to explore the quantum regime existing at the beginning of time – the Big Bang. The reasonableness of its results suggests that we are on the right track for understanding the Big Bang Epoch.

10.8 Form of Hubble's Law in the Generalized Robertson-Walker Model

Our generalized Robertson-Walker metric assumes an inhomogeneous space with some fixed center at $\v{r} = 0$. Presumably this center was the point at which the Big Bang took place at the beginning of the universe. In this section see how Hubble's Law emerges in the generalized model.

Hubble's law is one of the cornerstones of modern cosmology. While one might think a scale factor that depended on the radius coordinate might not be consistent with Hubble's Law it is easy to show that Hubble's Law is satisfied provided that the scale factor factorizes as required by the tr-component of Einstein's equations (eq. 10.4.14) for our generalized Robertson-Walker model.

First we give a simple derivation of Hubble's law for the case of a separable scale factor:

$$A(t, \v{r}) = a(t)b(\v{r}) \tag{10.4.12}$$

under the assumption that some remote galaxy lies on the same radial line as the line from the origin of the space coordinates to our galaxy. The proper distance between the remote galaxy and our galaxy has the form

$$D(t) = D_0 A(t, \v{r}) \tag{10.8.1}$$

The rate of recession of the remote galaxy is then

$$v = dD/dt = D_0 b(\v{r})da(t)/dt = HD(t) \tag{10.8.2}$$

with

$$H = d(\ln a(t))/dt \tag{10.8.3}$$

H is the Hubble Constant. If the speed of recession is small then it determines the first-order Doppler shift. If we denote the wavelength of the received radiation as λ_r and the wavelength of the radiation at the source as λ_s then the shift z is

$$z = \lambda_r/\lambda_s - 1 = v/c = HD/c \qquad (10.8.4)$$

Eq. 10.8.4 is Hubble's Law: the red shift of a galaxy is proportional to its distance. Note Hubble's Law in the generalized case follows from the factorization of $A(t, \check{r})$ with $a(t)$ satisfying the same differential equation as the Robertson-Walker scale factor.

Since a change of radial coordinate reduces the classical generalized model to the Robertson-Walker model, Hubble's Law can be proven in the general case of the non-collinearity of the coordinate origin, source and reception points.

11. Extended Robertson-Walker Model with Quantum Coordinates

In this chapter we introduce a form of quantum coordinates (called Two-Tier coordinates) into the Extended Robertson-Walker Model that we created in chapter 10. Its role is to eliminate the infinities at the Big Bang point. It has the remarkable feature that its effect becomes negligible shortly (within microseconds) after the Big Bang. Thus the form of the expansion scale factor is virtually identical to that of the usual Robertson-Walker scale factor. In Blaha (2018e) we showed there are a number of other important reasons for introducing our Quantum Coordinates beyond the problems with the Big Bang:[51]

"Originally Two-Tier coordinates were developed by this author to remove infinities that appear in perturbation theory calculations. We showed that the quantum smeared coordinates of Two-Tier Quantum Field Theory succeeded in removing all ultra-violet infinities in perturbation theory including the fermion triangle infinities. Remarkably the high precision, low energy[52] predictions of QED remained true in Two-Tier QED and thus remained consistent with experiment to a hitherto unsurpassed level of accuracy. 'Low' energy predictions in other quantum field theories also remained unchanged. At high energies, Two-Tier perturbation theory results are finite and consequently all ultra-violet infinities, to any order in perturbation theory, in *any number of space-time dimensions* were eliminated.

In addition to removing perturbation theory infinities Two-Tier coordinates enable us to define finite theories of Quantum Gravity and 'non-renormalizable' quantum field theories based on polynomial lagrangians, to tame vacuum fluctuations, to eliminate infinities associated with the Big Bang, and possibly to generate the explosive growth of the universe in its role as Dark Energy.[53]

---

[51] Extract from Blaha (2018e).
[52] Relative to a mass scale that was perhaps of the order of the Planck mass.
[53] See Blaha (2017b) and earlier books for details. This section is basically a summary of some features.

Two-Tier Quantum Field Theory is established on the most fundamental level.

…

Thus using Two-Tier Quantum Field Theory we can perform perturbation theory calculations that always yield a finite result in any number of dimensions.[54] This is not true if conventional Quantum Field is used.[55]"

In Blaha (2018e) we derived the form of the Unified SuperStandard Theory. It includes Gravitation, Dark Matter, a fourth fermion generation, and four layers of four generation fermions totaling 192 fermions. Similarly we found 192 non-abelian vector interaction bosons as well as a U(4) General Relativity-based non-abelian vector boson and a U(192) interaction rotations group.

We introduced a universal field Y^μ that eliminated the divergences that appear in perturbation theory without the use of a renormalization program. We now turn to the extension of our theory to the Cosmology of our universe, and the extension of Nature to include other universes, which together with our universe, reside in a D dimensional Megaverse.

It seemed appropriate to study the evolution of our universe from its beginning, and view it as a model of the general evolution of universes in the Megaverse. There does not appear to be any reason to view our universe as unique. Given the oft-noted tendency of Nature to repeat the characteristics of phenomenon we believe our universe should provisionally be viewed as a 'typical' universe.

Almost all of this chapter,[56] and the following chapters, first appeared in Blaha (2004) to derive our Quantum Big Bang Theory. The new aspect is its unification with a Steady State Theory of the Hoyle-Narlikar type to create the Hybrid Big_Bang-Steady_State Theory. Unlike Hoyle and Narlikar we can attribute the appearance of

---

[54] In particular, the fermion triangle divergence (anomaly) does not occur in our Two Tier Quantum Field Theory of the fermion sector. Thus there is no requirement for axion-like particles in the Megaverse (or in universes) although the possible existence of this type of particle is not ruled out.

[55] Blaha (2005a) provides a complete discussion of Two-Tier Quantum Field Theory.

[56] While the calculation is unchanged we have added clarifications of the original that might have led to misunderstandings particularly in clarifying the *constant* α vs. the scale function a(t).

mass-energy in the universe to an influx from the Megaverse using our surface tension force concept.

11.1 Dark Energy

A form of Dark Energy surfaced in Cosmology when it was noticed that the speed of expansion of our universe was expanding due to an unknown, and "undetectable" energy, dubbed Dark Energy. A theoretical framework was developed by A. Guth and others based on an *inflation* that provided a scenario for the expansion of the universe through an unidentified source called Dark Energy.

We shall now show that one cause of inflatons is the $Y^\mu(y)$ quantum field of our Extended Standard Model that also resolved QFT divergence problems. We will see $Y^\mu(y)$ also makes the Big Bang finite—no singularities; as well as freeing The Standard Model and Quantum Gravity[57] of infinities. *Thus $Y^\mu(y)$ has a remarkable triple role in our view – to eliminate the Big Bang singularity, to generate the explosive growth of the universe, and to remove infinities from The Standard Model and Quantum Gravity.* This happy coincidence of solutions reflects an Ockham's Razor approach to find the simplest general solution and Leibniz's Minimax Principle to find the most minimal solution that has maximal effects – two sides of the same coin in a sense.

Since the $Y^\mu(x)$ quantum gauge field is a free field (neglecting gravity) the initial state of the universe can be permeated with quanta of this field as well as particle quanta. We call the total energy of the free $Y^\mu(x)$ field within the universe Y Dark Energy.

11.2 The Big Bang Experimentally

In the light of our progress since 2004, we now provide a revised version that is quite similar to Blaha (2004).

The current state of our knowledge of the evolution of the universe has now been extended back in time to about 380,000 years after the Big Bang through recent astrophysical research. While this progress is encouraging we still face major issues: the

---

[57] See Blaha (2011c) and Blaha (2005a) for the removal of infinities in Quantum Gravity.

nature of Dark Matter (hopefully resolved by the Unified SuperStandard Theory), the nature and origin of Dark Energy (hopefully resolved in this chapter), and the events of those critical years before the 350,000 year point that we are slowly reaching experimentally. Those early years and the Big Bang itself were mysteries. This situation was especially critical since the early years of the universe apparently contain an uncertain beginning and an explosive growth.

In this chapter we will attempt to understand that unknown period in the neighborhood of t = 0 where we believe quantum coordinate effects play a major role. We will suggest that the inflationary growth of the universe, which is attributed to an unknown "particle", actually is caused by the energy of the q-number part of coordinates – the quantum field $Y^\mu(x)$ that we saw in earlier chapters.

11.3 The State of the Universe at t $= 0$

If we simply extrapolate the currently popular Standard Cosmological Models (with or without inflations) back to the Big Bang t = 0, we find a universe beginning as a single "mathematical point" with infinite mass density and infinite temperature. The Robertson-Walker metric scale factor $a(t) \equiv R(t)k^{1/2}$, which is a solution of the Einstein equation

$$\dot{a}^2 - 8\pi G\rho_{tot}a^2/3 = -k \qquad (11.1)$$

typically is solved for a perfect fluid under the assumption of a matter-dominated, or a radiation-dominated, phase of the universe. If we assume the universe is matter-dominated, then the energy density is

Matter-Dominated: $\qquad \rho = \rho_0/a(t)^3 \qquad\qquad (11.2)$

Under the alternate assumption that the universe is radiation-dominated we have[58]

Radiation-Dominated: $\qquad \rho = \rho_0/a(t)^4 \qquad\qquad (11.3)$

---

[58] *When we consider the mass-energy influx from the Megaverse we will assume the influx is primarily radiation just like in our universe.*

With either assumption we find that the scale factor behaves as

$$a(t) \propto t^n \tag{11.4}$$

where $0 < n < 1$. Thus $a(0) = 0$, and the universe reduces to a point with infinite density (eqns. 11.2 and 11.3), and with infinite temperature since

$$T \propto a^{-1}(t) \tag{11.5}$$

There are evidently grave difficulties in extrapolating the Standard Cosmological Model, or its current variants, to $t = 0$. The difficulty is compounded by the inherently quantum mechanical aspects that are normally associated with gravitation at ultra-small distances.

11.3.1 Two-Tier Quantum Coordinates in a Quantum Gravity

Currently, the only viable complete theory of Quantum Gravity is the Two-Tier Quantum Gravity of Blaha (2004) and (2005a). Blaha's type of quantum field theory has the interesting feature that all forces (particle propagators) become zero at very short distances (presumably much less than the Planck mass). Thus a point universe could have an infinite density of essentially "non-interacting" matter as a quasi-stable state. Furthermore if one uses the Extended Robertson-Walker metric as described in Blaha (2004), and Chapter 10 of this book, one finds a classical scale factor of the form:

$$A(t, \check{r}) = a(t)b(\check{r}) \tag{11.6}$$

where $a(t)$ satisfies eq. 11.1.

We now assume $A(t, \check{r})$ is a quantum operator due to the Two-Tier quantization of $\check{r}$ as shown in chapter 7. *The relation of the R scale factor, and other forms of the scale factor, is summarized by*

$$R(t) \rightarrow A_0(t, r) \tag{10.2.1}$$
$$A_0(t, r) = a(t)b_0(r) \tag{10.4.6}$$
$$A \equiv A(t, \check{r}) = a(t)b(\check{r}) \tag{10.4.12}$$

Since quantum effects will play a role near t = 0 (the "Big Bang") it is likely that the expectation value of $A(t, \check{r})$ will have the form

$$<A(t, \check{r})> = <a(t)><b(\check{r}, t)> \tag{11.7}$$

where

$$<b(\check{r}, t)> \rightarrow \beta(\check{r})/<a(t)> \tag{11.8}$$

as t → 0. Then the zero of $<a(t)>$ might be cancelled with the result

$$<A(t, \check{r})> \rightarrow \beta(\check{r}) \neq 0 \tag{11.9}$$

Quantum effects would thus eliminate the singularities at t = 0. A quantized version of the generalized Robertson-Walker model[59] opens the possibility of a universe with a finite size, density, and temperature at the time of the Big Bang.

With that possibility in mind, we will use an extension of Blaha (2004)'s Two-Tier theory to develop an enhanced version of his quantum model of the universe in the neighborhood of t = 0. Starting from eq. 11.7.1 of Blaha (2004) which is eq. 10.7.1 in Chapter 10:

$$A(t, \check{r}) = a(t)b(\check{r}) = 2\alpha k^{-\frac{1}{2}}a(t)[1 + \alpha^2 \check{r}^2]^{-1} \tag{11.7.1}$$

where α is a constant[60] specified by eq. 10.6.3m we introduce a Two-Tier variable Y^μ (as in chapter 7 with the identification[61]

$$\check{r} \equiv M_c X = M_c(y + iY/M_c^2)$$

where Y = |**Y**| we see

$$b(\check{r}) = b(y, t) = 2\alpha k^{-\frac{1}{2}}[1 + \alpha^2(M_c y + iY/M_c)^2]^{-1} \tag{11.10}$$

---

[59] This model appears in Blaha (2004).
[60] The constsant α is not time dependent; a(t) is time dependent. The lack of an argument further distinguishes α from a(t).
[61] α is a constant and not the fine structure constant.

If (as seen later)

$$Y = -M_c[a_1(y, t) - a_2(y, t)a(t)]^{\frac{1}{2}}/a + iM_c^2 a_3(y, t) \qquad (11.11)$$

and if, as $t \to 0$,

$$a_1(y, t) \to a_1(y, 0) = 1 \qquad (11.12)$$
$$a_2(y, t) \to a_2(y, 0) \neq 0 \qquad (11.13)$$
$$a_3(y, t) \to a_3(y, 0) = y \qquad (11.14)$$

then we find

$$b(y, t) \to 2ak^{-\frac{1}{2}}/[a_2(y, 0)a(t)] \qquad (11.15)$$

and

$$<A(t, \check{r})> = <a(t)b(\check{r})? = 2ak^{-\frac{1}{2}}/a_2(y, 0) + a(t)\beta_1(y) + a^2(t)\beta_2(y) + \ldots \qquad (11.16)$$

$$\to 2ak^{-\frac{1}{2}}/a_2(y, 0) \qquad \text{as} \quad t \to 0$$

Thus, under these circumstances, space does not collapse to a point, and the density and temperature—as well as other parameters of interest—are finite. *Yet the features of the Standard Cosmological Model at larger times still remain valid.*

11.3.2 Assumptions of Our Two-Tier Quantum Theory

In our model we will make the following assumptions about the universe near $t = 0$:

1. The particles in the universe consist of fundamental elementary particles – gravitons, photons, electrons, neutrinos, quarks, gluons and so on – and their corresponding anti-particles.

2. The particles are described by Two-Tier quantum field theory. In this type of quantum field theory all particle interactions become negligible at very short distances (as described in chapter 7 of Blaha (2004)) and so the forces

between particles may be neglected near t = 0 when the universe is immensely *small*.

3. The energy of the universe can be viewed as consisting of particles – bosons and fermions – each species having blackbody energy distributions since the universe is the best of all possible black bodies.

4. The enormous energy of the universe, even if confined to a small region, makes the classical Einstein equations a good approximation *due to its macroscopic nature* with one proviso (item 5). Therefore we assume the Extended Robertson-Walker metric of Blaha (2004) and chapter 10.

5. In the neighborhood of t = 0 when the universe is effectively confined to a region whose scale is set by the Planck mass or smaller the quantum nature of the Two-Tier coordinate X^μ becomes significant. In particular the Y^μ field causes a profound change in the behavior of the scale factor A(t, r) as t → 0. (Note: $X^\mu(y) = y^\mu + iY^\mu(y)/M^2$ defines the quantum coordinates where y^μ is a c-number coordinate and Y^μ a free q-number field similar to the electromagnetic field.)

6. The Y^μ quanta are assumed to have a black body spectrum[62] – just like elementary particles – reflecting their continuous emission and absorption by gravitons and other elementary particles. The Y^μ blackbody spectrum is implemented via a coherent state. Effectively the coherent state opens a small "bubble" into complex space-time changing the dynamics of the universe at t = 0.

---

[62] This is the only reasonable choice for the spectrum is a black body spectrum given the confinement of the field to the ultimate black body – the universe.

7. We will calculate the expectation value of the quantum field operator Y^μ in a closed Robertson-Walker space. In principle we must use the Extended Robertson-Walker metric since the scale factor will depend on both r and t through its dependence on the expectation value of Y^μ.

11.4 Two-Tier Quantum Hybrid Model for the Beginning of the Universe

Our approach in this, and the following, sections will follow a modest program using the known theoretical foundations of elementary particle physics: the Extended Standard Model unified with Quantum Gravity in a Two-Tier quantum field theoretic framework. We will supplement this framework with natural assumptions about the initial conditions of the universe in order to develop a theory describing the evolution of the universe from its initial state.

11.4.1 Einstein Equations Near t = 0

There is no physical reason to believe that the universe at the beginning of time, t = 0, was a mathematical point of infinite temperature and density since the extrapolation of the scale factor of the Standard Cosmological Model to t = 0 is unwarranted for many reasons including quantum considerations.

Ideally we would use the Quantum Theory of Gravity to establish the physical theory of the universe near t = 0. However a quantum calculation of the global structure of the universe near t = 0 is not feasible. In view of this situation we must find an approximation that captures the physics of the universe near the Big Bang. One approach is based on the macroscopic energy of the early universe. One can expect that a classical gravitation model with appropriate quantum corrections may be a reasonable approximation to the early state of the universe. After all, macroscopic bodies are described by classical physics in general. And the universe is a macroscopic body by virtue of its content at the point of the Big Bang despite its small size. Therefore we will assume that we may start with a classical gravitation model and then introduce quantum effects.

The natural first choices – based on symmetry considerations – are a Robertson-Walker model and the Extended Robertson-Walker model described in chapter 8 of Blaha (2004) and Chapter 10 here. The quantum part, that we will shortly introduce, will require us to use an Extended Robertson-Walker model since quantum corrections reduce the symmetry to a maximally symmetric *two-dimensional* subspace within a four-dimensional space-time. *The quantum part eliminates the equivalence of the classical Robertson-Walker model and the Extended Robertson-Walker models* that was described in section 8.7 of Blaha (2004).

Therefore we begin with the classical c-number equation for the invariant interval defined by

$$d\tau^2 = dt^2 - A^2(t, \check{r})[d\check{r}^2 + \check{r}^2(d\theta^2 + \sin^2\theta\, d\varphi^2)] \tag{11.17}$$

where

$$A(t, \check{r}) = a(t)b(\check{r}) = 2\alpha k^{-\frac{1}{2}}a(t)[1 + \alpha^2\,\check{r}^2]^{-1} \tag{11.18}$$

is a q-number and where α is a constant[63] given by eq. 10.6.3. The scale factor $a(t)$ is the solution of the Einstein equation:

$$\dot{a}^2(t) - 8\pi G\rho_{tot}a^2(t)/3 = -k \tag{11.19}$$

Next we introduce quantum coordinates

$$X^\mu = y^\mu + i\,Y^\mu(y)/M_c^2 \tag{11.20}$$

We choose the same transverse gauge for Y^μ as we did in chapter 7 of Blaha (2004):

$$\partial Y^i/\partial y^i = 0 \tag{11.21}$$

$$Y^0 = 0 \tag{11.22}$$

As a result we make the identification (definition of coordinates)

---

[63] We use α to denote a constant, later set equal to 1, and $a(t)$ to denote the solution of the Einstein equation above. They are not connected to each other.

$$X^0 = y^0 \equiv t \tag{11.23}$$
$$X^j = y^j + i\, Y^j(y)/M_c^2 \equiv M_c^{-1}\check{r}^j \tag{11.24}$$

The mass factor on the right side of the equal sign in eq. 11.20 is required on dimensional grounds if y is to have the usual dimension of length (inverse mass). As a result, since $\check{r} \in [0, 1]$ by Blaha (2004), eq. 11.24 implies

$$y = |\mathbf{y}| \in [0, M_c^{-1}] \tag{11.25}$$

There are two constants with the dimension of mass to a power: k and M_c. The constant k determines the curvature of space – a large-scale feature of Robertson-Walker models. The constant M_c is related to the very short distance behavior of the theory – high energy phenomena with energies of the order of the Planck mass or larger, and, as we will see, the origin of the universe – a short distance, high energy phenomena as well. Therefore we have also chosen to use M_c on the right side of eq. 11.24.

Since X^0 is a c-number and since the density $\rho(t)$ is a large c-number to very good approximation we will assume a(t) is the c-number solution of the classical c-number eq. 11.19 as $t \to 0$. Further we assume that quantum effects appear solely through b($\check{r}$). We also assume that the q-number equivalent of the c-number b($\check{r}$) is b(M_cX). The q-number function b(M_cX) satisfies the functional equation:[64]

$$k + (M_c^2 X b^2)^{-1}\, \partial(Xb'/b)/\partial X = 0 \tag{11.26}$$

where

$$b' = \partial b/\partial X \quad b = b(M_cX) \tag{11.27}$$

and $X = (\mathbf{X}\cdot\mathbf{X})^{1/2}$. The formal solution of eq. 11.26 has the same functional form as the c-number solution b($\check{r}$) in eq. 10.5.4. Therefore

$$b(M_cX) = :2ak^{-1/2}[1 + a^2 M_c^2 X^2]^{-1}: \tag{11.28}$$

---

[64] See eq. 10.5.3 for the c-number equivalent.

where α is the constant specified by eq. 10.6.3 and where we have specified normal ordering with colons **: ... :** to avoid trivial divergences.

Since eq. 11.28 is a q-number expression we must find the scale factor as the expectation value of A(t, X) for a suitable state. We note that, at this point, the invariant interval is an operator expression of the form:

$$d\tau^2 = dt^2 - B^2(t, X)[dX^2 + X^2(d\theta^2 + \sin^2\theta \, d\varphi^2)] \qquad (11.29)$$

where

$$B(t, X) = M_c A(t, M_c X) = a(t)b_M(X) \qquad (11.30)$$

and

$$b_M(X) = M_c b(M_c X) = :2aM_c k^{-\frac{1}{2}}[1 + a^2 M_c^2 X^2]^{-1}: \qquad (11.31)$$

where α is a constant (eq. 10.6.3). Thus the expectation value of $b_M(X)$ also must be calculated in order to determine the invariant interval's expectation value.

11.4.2 Y Black-Body Coherent States

The Y quanta are continuously being emitted and absorbed by the particles in the primeval universe. As such, they may be expected to have a blackbody energy spectrum that is similar to that of the particles from which they derive their existence. In particular one expects the temperature T associated with their blackbody energy distribution to be the same as that of the "real" particles in the universe. After all, the universe is a type of black body.

Thus the blackbody energy of Y-quanta as a function of frequency v per unit volume per unit frequency is assumed to be:

$$u_v = 8\pi hc^{-2}v^3 \, [e^{hv/\kappa T} - 1]^{-1} \qquad (11.32)$$

where c is the speed of light, h is Planck's constant, and κ is Boltzmann's constant. At this point we adopt units in which c = 1 and $\hbar$ = 1.

The Hamiltonian for the Y field has a form that is familiar from electrodynamics

$$H = \int d^3y \; \mathscr{H}_Y(y) = \tfrac{1}{2}\int d^3y \; :E_Y^2 + B_Y^2: \; = \int d^3p \; \omega \sum_\lambda a^\dagger(p,\lambda)a(p,\lambda) \qquad (11.33)$$

where E_Y and B_Y are the "electric" and "magnetic" parts of the Y field, $\omega = p^0 = |\vec{p}| = 2\pi v$ is the energy (in our units), and λ labels the polarization. Note that we are using "infinite volume" continuum quantization formulation.

We now define coherent Y field bra and ket states that yield a spherically symmetric blackbody distribution as the eigenvalue of the Hamiltonian H:

$$|BB, T> = N \, \exp[\int d^3p \; f(\omega,T)\sum_\lambda a^\dagger(p, \lambda)]|0> \qquad (11.34)$$

$$<BB, T| = N^* <0|\exp[\int d^3p \; f^*(\omega,T)\sum_\lambda a(p, \lambda)] \qquad (11.35)$$

where $\omega = |\vec{p}|$, and where N is a normalization factor. The expectation value of H is

$$<BB, T|H|BB, T> = \int d^3p \; 2\omega|f(\omega,T)|^2 \qquad (11.36)$$
$$= \int d\omega \; 8\pi\omega^3|f(\omega,T)|^2 \qquad (11.37)$$
$$= \int dv \; 16\pi^2\omega^3|f(\omega,T)|^2 \qquad (11.38)$$

where $\omega = 2\pi v$ in our units ($c = 1$, $\hbar = 1$), and where the factor of two in eq. 11.36 is the number of polarizations.

The expectation value (eigenvalue) of the energy per unit frequency is

$$H_v = 16\pi^2\omega^3|f(\omega,T)|^2 \qquad (11.39)$$

We relate H_v to the blackbody energy *per unit volume* per unit frequency u_v using

$$H_v = u_v(2\pi/\omega)^3 \qquad (11.40)$$

where the factor of $(2\pi/\omega)^3$ makes the right side of eq. 11.40 the blackbody energy per unit frequency in the continuum case of a quantum field in a space of infinite volume. Thus we find

$$f(\omega,T) = \omega^{-3/2}[e^{\omega/\kappa T} - 1]^{-\frac{1}{2}} \tag{11.41}$$

with the phase of $f(\omega,T)$ set to zero.

11.4.3 Expectation Value of Y in Coherent States

As a preliminary to the evaluation of the operator scale factor in eq. 11.31 we will evaluate the expectation value of powers of the Y field between black body coherent states defined by eqns. 11.34-11.35. We will then determine the expectation value of $b_M(X)$ to obtain the behavior of the overall scale factor near t = 0. It should be apparent to the reader that the expectation value of $b_M(X)$ is dependent on t as well as y due to the time dependence of the Y field. Thus the scale factor will exhibit a considerably more intricate behavior than simply its a(t) dependence.

The Fourier expansion of the Y field is:

$$Y^i(z) = \int d^3p \, N_0(\omega) \sum_{\lambda=1}^{2} \varepsilon^i(p, \lambda)[a(p,\lambda) \, e^{-ip \cdot z} + a^\dagger(p,\lambda) \, e^{ip \cdot z}] \tag{11.42}$$

for i = 1, 2, 3 where z^μ will be set equal to y^μ later, and where

$$N_0(\omega) = [(2\pi)^3 2\omega]^{-\frac{1}{2}} \tag{11.43}$$

and

$$\omega = (\mathbf{p}^2)^{\frac{1}{2}} = p^0 \tag{11.44}$$

with $\vec{\varepsilon}(p, \lambda)$ being the polarization unit vectors for $\lambda = 1, 2$ and $\eta_{\mu\nu}p^\mu p^\nu = 0$. The expectation value of Y between the |BB, T> states is:

$$<BB, T|Y^i(z)|BB, T> = \int d^3p \, N_0(\omega)f(\omega,T)[e^{-ip \cdot z} + e^{ip \cdot z}] \sum \varepsilon^i(p, \lambda) \tag{11.45}$$

The evaluation of eq. 11.45 (and spherical symmetry) gives

$$<BB, T|Y^i(z)|BB, T> = \hat{y}^i \int d^3p \, N_0(\omega)f(\omega,T)[e^{-ip \cdot z} + e^{ip \cdot z}] \sum_\lambda \hat{\mathbf{z}} \cdot \varepsilon(p, \lambda) \tag{11.46}$$

$$\equiv \hat{\mathbf{z}}^{i}\, Y_{BB}(t,\, z)$$

where $\hat{\mathbf{z}} = \vec{\mathbf{z}}/|\vec{\mathbf{z}}|$ is the unit 3-vector in the direction of $\vec{\mathbf{z}}$, $z = |\vec{\mathbf{z}}|$, and $p{\cdot}z = \omega(t - z\cos\theta)$. We define a spatial coordinate system – choosing the z-axis parallel to $\vec{z}$. Then we have

$$\vec{z} = (0,\, 0,\, z) \tag{11.47}$$
$$\vec{p} = (\sin\theta\cos\phi,\, \sin\theta\sin\phi,\, \cos\theta) \tag{11.48}$$
$$\vec{\varepsilon}(p,1) = (\cos\theta\cos\phi,\, \cos\theta\sin\phi,\, -\sin\theta) \tag{11.49}$$
$$\vec{\varepsilon}(p,2) = (-\sin\phi,\, \cos\phi,\, 0) \tag{11.50}$$

with the result (taking account of eq. 11.46)

$$Y_{BB}(t,\, z) = {<}BB,\, T|\,\hat{\mathbf{z}}{\cdot}\,\mathbf{Y}(t,\, z)\,|BB,\, T{>}$$
$$= 2\pi\!\int_{0}^{\infty}\! d\omega\, \omega^{2} N_{0}(\omega) f(\omega, T)\!\int_{0}^{\pi}\! d\theta\, \sin^{2}\theta\, [e^{-ip{\cdot}z} + e^{ip{\cdot}z}] \tag{11.51}$$

where $p{\cdot}z = \omega(t - z\cos\theta)$ with $z = |\vec{z}|$. We will develop integral representations and approximations to Y_{BB} in a later section.

11.4.4 Expectation Values of the Scale Factor A(t, X) and the Invariant Interval $d\tau^2$

The scale factor

$$b_{M}(X) = :2aM_{c}k^{-\frac{1}{2}}[1 + \alpha^{2}M_{c}^{2}X^{2}]^{-1}: \tag{11.52}$$

where α is a constant given by eq. 10.6.3, is a normal-ordered q-number expression. We can formally expand (define) this expression as a power series of normal-ordered powers of X^{2} and then evaluate it between blackbody coherent states. First we note that

$$<BB, T|:Y^{i_1}(z)Y^{i_2}(z)Y^{i_3}(z)Y^{i_4}(z) \ldots Y^{i_n}(z):|BB, T> = \hat{z}^{i_1}\hat{z}^{i_2}\hat{z}^{i_3}\hat{z}^{i_4} \ldots \hat{z}^{i_n}(Y_{BB}(t, z))^n$$

$$(11.53)$$

We now set $z^\mu = y^\mu$, and use Y_{BB} to represent $Y_{BB}(t, \mathbf{y})$:

$$Y_{BB} \equiv Y_{BB}(t, \vec{y}) \qquad (11.54)$$

Thus

$$b_{BB}(y, t) = <BB, T|b_M(X)|BB, T>$$

$$= 2aM_c k^{-\frac{1}{2}}\{1 + a^2[M_c^2 y^2 + 2i\, y Y_{BB} - Y_{BB}^2/M_c^2]\}^{-1} \qquad (11.55)$$

and

$$B_{BB}(t, y) = <BB, T|B(t, X)|BB, T> = a(t)b_{BB}(y, t) \qquad (11.56)$$

by eq. 11.29-31 where α is a constant (eq. 10.6.3). The expectation value of the q-number invariant interval (eq. 11.29) is the c-number expression:

$$d\tau_{BB}^2 = <BB, T|d\tau^2|BB, T>$$

$$= dt^2 - B_{BB}^2(t, y)[dX_{BB}^2 + X_{BB}^2(d\theta^2 + \sin^2\theta\, d\varphi^2)] \qquad (11.57)$$

where

$$X_{BB} = y + iY_{BB}/M_c^2 \qquad (11.58)$$

and

$$dX_{BB} = dy(1 + iM_c^{-2}\partial Y_{BB}/\partial y) \qquad (11.59)$$

The appearance of Y_{BB} in the expression for the invariant interval (eq. 11.57) has two effects: it introduces complex space-time into the model and the Extended Robertson-Walker metric is no longer equivalent to the Robertson-Walker metric as it would be in the classical case.

We will see that Y_{BB} approaches zero at large times thus leading to the conventional Robertson-Walker models. But at small times of the order of the Planck time near the Big Bang we enter a brave new world of complex space-time. We will investigate the nature of this new complex world in the succeeding sections.

11.4.5 Representation and Approximations for $Y_{BB}(t, z)$

The angle integral in eq. 11.51 can be performed to yield

$$Y_{BB}(t, z) = \pi^{\frac{1}{2}}z^{-1} \int_0^\infty d\omega \, \omega^{-1}(e^{\omega/\kappa T} - 1)^{-\frac{1}{2}}\cos(\omega t)J_1(\omega z) \qquad (11.60)$$

where $J_1(\omega z)$ is a Bessel function using 3.915.5 of Gradshteyn (1965).

11.4.5.1 Some Representations of Y_{BB}

The integral in eq. 11.60 does not appear to be simply expressible in terms of standard transcendental functions. A series representation of the integral can be obtained by expanding the exponential factor due to the Planck distribution:

$$Y_{BB}(t,z) = \tfrac{1}{2}\pi^{\frac{1}{2}}\kappa T \sum_{n=0}^{\infty} (2n)![2^{2n}(n!)^2]^{-1}\{(2n+1+2i\kappa Tt)^{-1}F(\tfrac{1}{2}, 1; 2; -[\kappa Tz/(n+\tfrac{1}{2}+i\kappa Tt)]^2) +$$
$$+ (2n+1 - 2i\kappa Tt)^{-1} F(\tfrac{1}{2}, 1; 2; -[\kappa Tz/(n+\tfrac{1}{2} - i\kappa Tt)]^2)\} \qquad (11.61)$$

where $F(a, b; c; w)$ is a hypergeometric function.[65]

Using an integral representation[66]

$$F(a, b; c; w) = \Gamma(c)[\,\Gamma(b)\Gamma(c - b)]^{-1} \int_0^1 dt \, t^{b-1}(1 - t)^{c-b-1}(1 - tw)^{-a}$$

for $F(\tfrac{1}{2}, 1; 2; w)$ we see eq. 11.61 can be written in terms of simpler algebraic expressions:

$$Y_{BB}(t, z) = \tfrac{1}{2}\pi^{\frac{1}{2}}(\kappa Tz^2)^{-1} \sum_{n=0}^{\infty}(2n)![2^{2n}(n!)^2]^{-1}\{[(n+\tfrac{1}{2} + i\kappa Tt)^2 + (\kappa Tz)^2]^{\frac{1}{2}} +$$
$$+ [(n+\tfrac{1}{2} - i\kappa Tt)^2 + (\kappa Tz)^2]^{\frac{1}{2}} - (2n+1)\} \qquad (11.62)$$

---

[65] Based on the integral 6.613.1 on p. 711 of Gradshteyn (1965).
[66] Magnus (1949) p. 8.

Eq. 11.62 shows the limit of $Y_{BB}(t, z)$ for large t is

$$Y_{BB}(t, z) \to \pi^{\frac{1}{2}}\kappa T \sum_{n=0}^{\infty}(2n)![2^{2n+2}(n!)^2]^{-1}(2n + 1)[(n + \tfrac{1}{2})^2 + (\kappa Tt)^2]^{-1} \to 0 \qquad (11.63)$$

if $tT \to \infty$ as $t \to \infty$ as we see in cosmological models (see section 11.4).

Thus

$$(b_{BB}(y, t))^2 dX^2 \to 2a M_c^2 k^{-1}[1 + a^2 M_c^2 y^2]^{-2} dy^2 \equiv 2ak^{-1}[1 + a^2 \check{r}^2]^{-2} d\check{r}^2 \qquad (11.64)$$

where α is a constant (eq. 10.6.3), for large t showing the Two-Tier cosmological model becomes a Robertson-Walker model at large times. However, the Two-Tier standard cosmological model is very different at small times of the order of the Planck time near the Big Bang point.

11.4.5.2 Approximate Solution for Y_{BB}

The integral representation and power series representation of $Y_{BB}(t, y)$ do not reveal the physical behavior of the model for small times t and distances y. Therefore we will examine an approximation for $Y_{BB}(t, y)$ for ranges of y, t and T that are relevant for our considerations. We begin by scaling the integration variable in eq. 11.60 with the result:

$$Y_{BB}(t, y) = \pi^{\frac{1}{2}}y^{-1} \int_0^{\infty} d\omega\, \omega^{-1}(e^{\omega} - 1)^{-\frac{1}{2}}\cos(\omega\kappa Tt)J_1(\omega\kappa Ty) \qquad (11.65)$$

The blackbody exponential factor $(e^{\omega} - 1)^{-\frac{1}{2}}$ in the integrand of $Y_{BB}(t, y)$ enables the leading order approximate behavior of $Y_{BB}(t, y)$ to be determined for $0 \le t \lessgtr 10^{108}$ s – for all time, practically speaking. In a later section (section 11.3.4) we will see that our approximation to the integral in eq. 11.65 is consistent with the solution that we obtain for Y_{BB}, for the scale factor and thus for the temperature T. The approximations that we will make in eq. 11.65 are

$$\cos(\omega\kappa Tt) \approx 1 \qquad (11.66)$$
$$J_1(\omega\kappa Ty) \approx \omega\kappa Ty/2 \qquad (11.67)$$

They are based on $\kappa Tt \ll 1$ and $\kappa Ty \ll 1$ for all y ($0 \le y \le M_c^{-1}$). The exponential factor tends to limit contributions to the integral to small ω. After making these approximations we find

$$Y_{BB}(t,\ y) \simeq \tfrac{1}{2}\ \pi^{\frac{1}{2}}\kappa T \int\limits_0^\infty d\omega\ (e^\omega - 1)^{-\frac{1}{2}}$$

$$\simeq \pi^{3/2}\kappa T/2 \tag{11.68}$$

The limit as t gets large can also be approximately determined from eq. 11.65. For large t such that κTy is small (and ω is small due to the exponential Planck distribution factor) we can again approximate the Bessel function with its leading power series expansion term and the exponential factor can again be approximated by $e^\omega - 1 \approx \omega$ so that eq. 11.65 becomes approximately

As $t \to \infty$:
$$Y_{BB}(t,\ y) \simeq \pi^{\frac{1}{2}}2^{-1}\ \kappa T \int\limits_0^\infty d\omega\ \omega^{-\frac{1}{2}}\cos(\omega\kappa Tt)$$

$$= \pi 2^{-3/2}[\kappa T/t]^{\frac{1}{2}} \tag{11.69}$$

using 3.751.2 of Gradshteyn (1965). We note that, while κTy is small, κTt could possibly have been large in either a matter-dominated or radiation-dominated universe since it grows as t to a positive power (see section 11.3.) *However, since $Y_{BB}(t,\ y)$ approaches zero for large times its impact can only be seen in the initial formative stages of the universe near t = 0.*

11.5 The Scale Factor a(t) Near t = 0

The "time factor" a(t) of the scale factor $B_{BB}(t,\ y)$ appears in

$$B_{BB}(t,\ y) = <BB,\ T|B(t,\ X)|BB,\ T> = a(t)b_{BB}(y,\ t) \tag{11.70}$$

and is determined by the classical Einstein equation:

$$\dot{a}^2(t) - 8\pi G\rho_{tot}a^2(t)/3 = -k \tag{11.71}$$

As we have argued earlier, the source determining a(t) for small times in the neighborhood of t = 0 (the time of the Big Bang) is a large, macroscopic, classical density $\rho_{tot}(t)$ and thus a(t) may be considered to be a c-number quantity determined by the c-number Einstein equation to good approximation. This approximation should continue to hold even if this macroscopic density becomes enormous as t → 0. The quantum effects near t = 0 in the Two-Tier model, that we have developed, appear in the factor $b_{BB}(y, t)$, which we evaluated in previous sections.

11.6 A Complex Blackbody Temperature Near t = 0

The blackbody temperature T for relativistic particles (presumably the dominant type of particles near t = 0) is inversely proportional to the scale factor. <u>At large times</u> the blackbody temperature has the form

$$T = T_0/a(t) \qquad\qquad (11.72)$$

where T_0 is a constant.

At time periods in the neighborhood of t = 0 (the Big Bang) space has three complex dimensions in the Two-Tier model. Temperature can be viewed as a measure of the root mean square speed (or the "average energy") of the components of the perfect fluid that we have assumed. In the case of a gas of particles of average energy E:

$$T = E/(3k/2) \qquad\qquad (11.73)$$

In a complex space it is quite natural for the root mean squared speed to be complex as well. As a result complex temperatures naturally follow. Thus we will define

$$T = T_0/B_{BB}(t, y) \qquad\qquad (11.74)$$

<u>for all time since t ≥ 0</u>. Since $B_{BB}(t, y)$ approaches $M_c a(t) b(\mathring{r})$ at large times we find its large time behavior is consistent with those of standard Robertson-Walker models. At times near t = 0, the blackbody temperature T is complex since space is complex and

complex kinetic energy is allowed. Then we apply a Complex Lorentz (Reality) group transformation to obtain the physical temperature.

In the case of the complex temperature T above, the corresponding physical temperature is its absolute value (obtained by multiplying T by a phase factor from the Reality group)

$$T_{physical} = |T_0/B_{BB}(t, y)| \qquad (11.74a)$$

We shall use eq. 11.74 for the temperature, transforming the results below, afterwards, to real values using the 4-dimensional Reality group.

11.7 The Nature of the Universe Near t = 0

At this point we are ready to examine the Two-Tier model for the Big Bang period.

11.7.1 Behavior of the Complete Scale Factor B(t, y) Near t = 0

The behavior of the expectation value of the scale factor B(t, y), under the assumption that the Y quanta have a blackbody spectrum, is described by the equations:

$$b_{BB}(y, t) = 2aM_c k^{-\frac{1}{2}} \{1 + a^2[M_c^2 y^2 + 2i\, y Y_{BB} - Y_{BB}^2/M_c^2]\}^{-1} \qquad (11.75)$$
$$B_{BB}(t, y) = <BB, T|B(t, X)|BB, T> = a(t)b_{BB}(y, t) \qquad (11.76)$$

where α is a constant (eq. 10.6.3), and

$$Y_{BB}(t, y) \simeq \pi^{3/2} \kappa T/2 \qquad (11.77)$$
$$a(t) = [2\pi G \rho_0 n^2/3]^{1/n}\, t^{2/n} \qquad (11.78)$$
$$T = T_0/B_{BB}(t, y) \qquad (11.79)$$

as t → 0. We will temporarily set the constant $\alpha = 1$ in the interests of simplicity knowing that this value results in a metric fully equivalent to the Robertson-Walker metric at large times. (Other values of a would also result in a metric equivalent to the Robertson-Walker metric at large times after a re-scaling of the radial coordinate.) Thus we may write

$$B_{BB}(t, y) \cong 2k^{-\frac{1}{2}}M_c a(t)\{1 + M_c^2 y^2 + iy\pi^{3/2}\kappa T_0/B_{BB} - \pi^3\kappa^2 T_0^2/(4M_c^2 B_{BB}^2)\}^{-1} \quad (11.80)$$

This quadratic algebraic equation for B_{BB} has the solutions:

$$B_{BB}(t, y) \cong (1 + M_c^2 y^2)^{-1}\{-i\chi M_c y + k^{-\frac{1}{2}}M_c a(t) \pm [\chi^2 - 2i\chi yk^{-\frac{1}{2}}M_c^2 a(t) + k^{-1}M_c^2 a^2(t)]^{\frac{1}{2}}\} \quad (11.81)$$

with

$$\chi = \pi^{3/2}\kappa T_0/(2M_c) \quad (11.82)$$

As t gets very large we obtain the equivalent of the Robertson-Walker metric scale factor in this approximation if we choose the plus sign in eq. 11.81 (assuming $a^2(t)$ becomes very large so other terms within the square root can be neglected):

$$B_{BB}(t, y) \rightarrow 2k^{-\frac{1}{2}}M_c a(t)/(1 + M_c^2 y^2) \quad (11.83)$$

Thus we must choose the plus sign in eq. 11.81:

$$B_{BB}(t, y) \cong (1 + M_c^2 y^2)^{-1}\{-i\chi M_c y + k^{-\frac{1}{2}}M_c a(t) + [\chi^2 - 2i\chi yk^{-\frac{1}{2}}M_c^2 a(t) + k^{-1}M_c^2 a^2(t)]^{\frac{1}{2}}\} \quad (11.84)$$

At t = 0 (the Big Bang) eq. 11.84 simplifies to (assuming a(0) = 0)

$$B_{BB}(0, y) \cong [\chi - i\chi M_c y]/(1 + M_c^2 y^2) \quad (11.85)$$

For small y, the real part of $B_{BB}(0, y)$ is a constant and the imaginary part of $B_{BB}(0, y)$ is proportional to y.

11.7.2 The Expectation Value of the Scale Factor $A_{BB}(0, y)$ near $t = 0$

Eq. 11.85 gives the approximate behavior of the expectation value of the scale factor $B_{BB}(0, y)$ near $t = 0$ as a function of y. If we compare eq. 11.85 with the mechanism described in eqns. 11.1.6 – 11.1.9 of section 11.1 for cancelling the a(t) factor within the complete scale factor we see that we have found the blackbody spectrum of the Y quanta implements this mechanism. The solution can be written in the form:

$$A_{BB}(t, y) = M_c^{-1}B_{BB}(t, y) \cong \beta_0(y) + \beta_1(y)a(t) + \ldots \qquad (11.86)$$

$$\beta_0(y) = x(1 - iM_cy)/[M_c(1 + M_c^2y^2)] \qquad (11.87)$$

$$\beta_1(y) = k^{-\frac{1}{2}}(1 - iM_cy)/(1 + M_c^2y^2) \qquad (11.88)$$

Eqns. 11.86–11.88 are expressed in terms of the y variable. They can be expressed in terms of ř as:

$$A(t, ř) = a(t)b(ř) = 2ak^{-\frac{1}{2}}a(t)[1 + a^2ř^2]^{-1} \qquad (11.89)$$

where α is set to $a = 1$. Evidently, we have $M_cy \equiv ř$ at the level of approximation that we are using. Furthermore we can use

$$ř = \{[1 - (1 - kr^2)^{\frac{1}{2}}]/[1 + (1 - kr^2)^{\frac{1}{2}}]\}^{\frac{1}{2}} \qquad (11.90)$$

to express ř in terms of the Roberson-Walker radial coordinate r. Thus we find that the Robertson-Walker scale factor a(t) becomes

$$a(t) \to (1 + M_c^2y^2)(2k^{-\frac{1}{2}})^{-1}A_{BB}(t, y) \equiv a_{BBRW}(t, ř) \qquad (11.91)$$

$$a_{BBRW}(t, ř) \cong \beta_{0RW}(ř) + \beta_{1RW}(ř)a(t) + \ldots \qquad (11.92)$$

using the subscript "RW" to denote quantities scaled to the standard Robertson-Walker metric, with

$$\beta_{0RW}(ř) = x(1 - iř)/[2k^{-\frac{1}{2}}M_c] \qquad (11.93)$$

$$\beta_{1RW}(\check{r}) = (1 - i\check{r})/2 \qquad\qquad (11.94)$$

where $\check{r}$ is specified by eq. 11.90.

Using the Reality group we can "rotate" the complex scale factor, which is a factor in coordinate expressions, to a real value—its absolute value

$$a(t) \rightarrow (1 + M_c^2 y^2)(2k^{-\frac{1}{2}})^{-1}|A_{BB}(t, y)| \equiv |a_{BBRW}(t, \check{r})| \qquad (11.91a)$$

We will study the implications of this scale factor in the following chapters.

12. Universe Mass-Energy Density

The previous chapters have examined the quantum part of the scale factor of the Quantum Extended Robertson-Walker Model. We found that the quantum scale factor had the form

$$A(t, ř) = a(t)b(ř, t) \tag{12.1}$$

where $a(t)$ is a c-number scale factor dependent on the total mass-energy density of the universe, and $b(ř, t)$ is a quantum scale factor dependent primarily on the second quantized Y^μ field (the imaginary part of the quantum coordinate X^μ.)

The expectation value of $A(t, ř)$ had the form

$$<A(t, ř)> = a(t)<b(ř, t)> \tag{11.7}$$

We evaluated the $<b(ř, t)>$ in chapter 10 under the assumption that Y quanta had a black body energy distribution. We continue the discussion in chapter 13.

13. From t = 0 to t = t_T: a(t) Scale Factor and Hubble Constant

13.1 Introduction

This chapter develops a numerical model for the scale factor of the Big Bang metastable state. Although the universe that we live in is almost flat according to recent WMAP experimental data, the small curvature of space closes the universe and has significant effects. Therefore we will not use a flat space approximation. Our goal is to obtain an order of magnitude understanding of the evolution of the universe from the beginning. Our calculated numerical quantities appear generally to be of the right order of magnitude. And the physical ideas appear to be consistent with a reasonable view of reality.

Chapter 11 describes the physical implications of quantum part b(ř, t) of the total scale factor A(ř, t). It depends on the black body Y quanta.

13.1 Total Mass-Energy Density from t = 0 to t = t_T

In this chapter we consider a(t), which we take to be a c-number quantity since it depends solely on macroscopic mass-energy densities as seen in

$$\rho_{tot}(t) = \rho_{crit}[\Omega_{\gamma}(t) + \Omega_{\Lambda}(t) + \Omega_m(t) + \Omega_d(t)] \equiv \rho_{crit}\Omega_H(t) \qquad (9.7)$$

The data that we use in this chapter and throughout are the combined results of the WMAP and SDSS data[67] based on the assumption of a non-flat space. In particular we use the following values:

h = Hubble parameter = 0.678(9)

---

[67] M. Tanabashi *et al* (Particle Data Group), Phys. Rev. D**98**, 030001 (2018).

$$\rho_{crit} = \text{Critical density} = 1.87840(9) \, h^2 \times 10^{-29} \, \text{g/cm}^{-3}$$
$$\Omega_\Lambda = \text{Dark Energy density}/\rho_{cr} = \rho_{de}/\rho_{cr} = 0.692 \pm 0.012$$
$$\Omega_d = \Omega_c = \text{Cold Dark matter density}/\rho_{cr} = \rho_c/\rho_{cr} = 0.1186(20) \, h^{-2}$$
$$\Omega_b = \text{Baryon density}/\rho_{cr} = \rho_b/\rho_{cr} = 0.02226 \, h^{-2}$$
$$\Omega_m = \Omega_b + \Omega_d$$
$$\quad = \text{pressureless Matter density}/\rho_{cr} = \rho_m/\rho_{cr} = 0.308 \pm 0.012$$
$$\Omega_{tot} = \Omega_m + \Omega_\Lambda = 1.000 \pm 0.024$$
$$t_0 = t_{now} = \text{Age of universe} = 13.80 \pm 0.04 \, \text{Gyr}$$

(13.1.1)

The time dependence of the scale parameter is given by the Einstein equation:

$$\dot{a}^2 - 8\pi G\rho_{tot}a^2/3 = -k \qquad (13.1.2)$$

. Also, based on an analysis of WMAP[68] data:

$$r_{universe}(t_{now}) = \text{visible radius of the universe} = 4.314 \times 10^{28} \, \text{cm} \qquad (13.1.3)$$

The reader is directed to the original papers for other parameters, error bars and a detailed analysis of the data. *We use units where h = c = 1 unless stated otherwise.*
We also use:[69]

$$\Omega_\gamma = \text{radiation density}/\rho_{cr} = \rho_\gamma/\rho_{cr} = 2.473h^{-2} \times 10^{-5} \, (T/2.7255)^4 h^{-2}$$
$$\quad = 5.38 \times 10^{-5} \qquad (13.1.4)$$

13.2 The Behavior of the Scale Factor a(t) after the Big Bang Period

The universe, as we know it today, contains a variety of forms of energy. The current densities of these forms of energy are listed in section 13.1. From them we can develop the form of the total energy density and project it back to the instants after the Big

---

[68] M. Tanabashi *et al* (Particle Data Group), Phys. Rev. D**98**, 030001 (2018).
[69] M. Tanabashi *et al op. cit.*

Bang. The Big Bang period is significantly different as we have seen in the preceding chapter.

The total mass-energy density as a function of time from eq. 9.6 neglecting $\Omega_T(t)$ is

$$\rho_{tot}(t) \equiv \rho_{crit}\Omega_{tot}(t) = \rho_{crit}[\Omega_\Gamma(t) + \Omega_M(t) + \Omega_\Lambda(t)]$$

$$= \rho_{crit}[\Omega_i(t) + \Omega_\Lambda(t) + \Omega_m(t)]$$

$$= \rho_{crit}[\Omega_\Gamma/a^4(t) + \Omega_M/a^3(t) + \Omega_\Lambda] \qquad (13.2.1)$$

where we treat the time dependence of Dark energy as similar to "normal" energy, and of Dark Matter as similar to "normal" matter (as in the Unified SuperStandard Theory.) based on well-known arguments.[70] Note

$$\Omega_M(t) = \Omega_m(t) = \Omega_b(t) + \Omega_d(t)$$

The time dependence of the scale parameter is given by the Einstein equation:

$$\dot{a}^2 - 8\pi G\rho_{tot}a^2/3 = -k \qquad (10.5.2)$$

where $a(t)$ is the Robertson-Walker scale factor with $a(t_{now}) = 1$.

Before proceeding to the solution of eq. 10.5.2 we need to obtain a reasonable estimate of the curvature constant k. We can use the Robertson-Walker expression for the radius of the universe

$$r_{universe}(t_{now}) = a(t_{now})/k^{\frac{1}{2}} \qquad (13.2.2)$$

or

$$k^{-\frac{1}{2}} = r_{universe}(t_{now}) \qquad (13.2.3)$$

Using this radius we obtain one estimate

$$k = 5.37 \times 10^{-58}\,cm^{-2} \qquad (13.2.4)$$

---

[70] Weinberg(1972), Dodelson(2003). We differ by not distinguishing Dark from normal based on our SuperStandard Theory.

13.2.1 General Form of a(t) Scale Factor Einstein Equation

We find the general form of the scale factor differential equation by combining eqs. 10.5.2 and 13.2.1

$$\dot{a}^2 - H_0^2 a^2(t)[\Omega_\Gamma/a^4(t) + \Omega_M/a^3(t)] = -k \qquad (13.2.1.1)$$

where the Hubble constant H_0 satisfies

$$H_0^2 = [d(\ln a)/dt]|_{t\,=\,t_{now}} = 8\pi G\rho_{cr}/3 \equiv 1.2 \times 10^{-56} h^2 \text{ cm}^{-2} \qquad (13.2.1.2a)$$

or

$$H_0 = 100 \text{ h km s}^{-1} \text{ Mpc}^{-1} = h \times (9.777752 \text{ Gyr})^{-1} =$$
$$= 1.1 \times 10^{-28} h \text{ cm}^{-1} \equiv 3.24 \times 10^{-18} h \text{ s}^{-1} \qquad (13.2.1.2b)$$

If we evaluate eq. 13.2.1.1 for the present time we find

$$k = [\Omega_\gamma + \Omega_\Lambda + \Omega_m]H_0^2 \cong [\Omega_m + \Omega_\Lambda]H_0^2 = H_0^2$$

$$= 5.56 \times 10^{57} \text{ cm}^{-2} \qquad (13.2.1.2c)$$

since

$$\Omega_m + \Omega_\Lambda \cong 1.00$$

We now define

$$\xi = k/H_0^2 \cong 1 \qquad (13.2.1.3)$$

for later use.

The solution of eq. 13.2.1.1 can be put into the form of integrals representing combinations of elliptic integrals:

$$\int_{a(t')}^{a(t)} da\ a\ H_0^{-1}\ [\Omega_\Lambda a^4 - \xi a^2 + \Omega_m a + \Omega_\gamma]^{-\frac{1}{2}} = \int_{t'}^{t} dt \qquad (13.2.1.4)$$

The result of these integrations is an implicit equation for a(t) that cannot be expressed in a simple closed form in terms of known functions. This equation can be easily solved numerically. A graph of a(t) is displayed in Fig. 13.2.1.1. Although it looks linear there are significant non-linearities in various parts of the plot of a(t). The radiation-dominated phase is not visible. It is a small slice of the plot since it amounts to less than 10^{13} s.

Because we know physically that approximations are possible for each of the various epochs: the matter-dominated epoch, the radiation-dominated epoch and so on, we can find physically meaningful approximations for each epoch. We therefore provisionally divide the life of the universe into two epochs: an explosive growth epoch, and an expanding epoch subdivided into matter-dominated and radiation-dominated phases.

We begin with the most recent phase, the explosive growth epoch, and then describe the previous phase, the expanding epoch, containing the matter dominated and radiation dominated phases. The results below in this chapter assume that $T > T_c$ implying no influx of Megaverse mass-energy. Chapter 14 considers the impact of a Megaverse mass-energy influx.

13.2.2 The Explosive Growth Epoch

The integral on the left in eq. 13.2.1.4 appears to support a simple approximation during the explosive growth phase. Notice that at $t = t_{now}$ we have $-\xi a^2 + \Omega_m a + \Omega_\gamma \cong -0.602$. The sum of these terms gets smaller as we proceed into the past. Therefore we approximate the left side with

$$\int_{a(t')}^{a(t)} da\, a H_0^{-1} [\Omega_\Lambda a^4]^{-\frac{1}{2}} = (H_0 \Omega_\Lambda^{\frac{1}{2}})^{-1} \int_{a(t')}^{a(t)} da/a = (H_0 \Omega_\Lambda^{\frac{1}{2}})^{-1} \ln[a(t)/a(t')] \cong t - t' \qquad (13.2.2.1)$$

or

$$a(t) = a(t') \exp[H_0 \Omega_\Lambda^{\frac{1}{2}}(t - t')] \qquad (13.2.2.2)$$

Thus we have deSitter-like exponential growth.[71] We would expect this growth phase to last approximately for the period where

---

[71] The Explosive Growth Epoch is also known as the **Steady State Cosmology**. It was pioneered by F. Hoyle, M.N.R.A.S. **108**, 372 (1948); F. Hoyle and J. V. Narlikar, Proc. R. Soc. London **A290**, 162 (1966); H. Bondi and T.

$$\Omega_\Lambda a^4(t_E) \approx \Omega_m a(t_E) \implies a(t_E) = .76 \qquad (13.2.2.3)$$

until the present where t_E is the time when $a(t_E) = .76$. Since we have normalized $a(t)$ by

$$a(t_{now}) = 1 \qquad (13.2.2.4)$$

we see that eq. 13.2.2.2 requires

$$a(t) = \exp[H_0\Omega_\Lambda^{\frac{1}{2}}(t - t_{now})] \qquad (13.2.2.5)$$

in the epoch that we have called the Explosive Growth Epoch. Note

$$H_0\Omega_\Lambda^{\frac{1}{2}} = 1.83 \times 10^{-18} \text{ s}^{-1} \qquad (13.2.2.6)$$

Gold, M.N.R.A.S. **108**, 252 (1948) and references therein. It implements a "perfect cosmological principle" by giving a universe that is the same at all points and times. It requires the continuous creation of matter with the origin of the created matter uncertain although it will be an enormous amount since it is distributed over the universe. In chapter 14 we consider a more definite theory in which "new" matter originates in the external Megaverse as a result of universe surface tension.

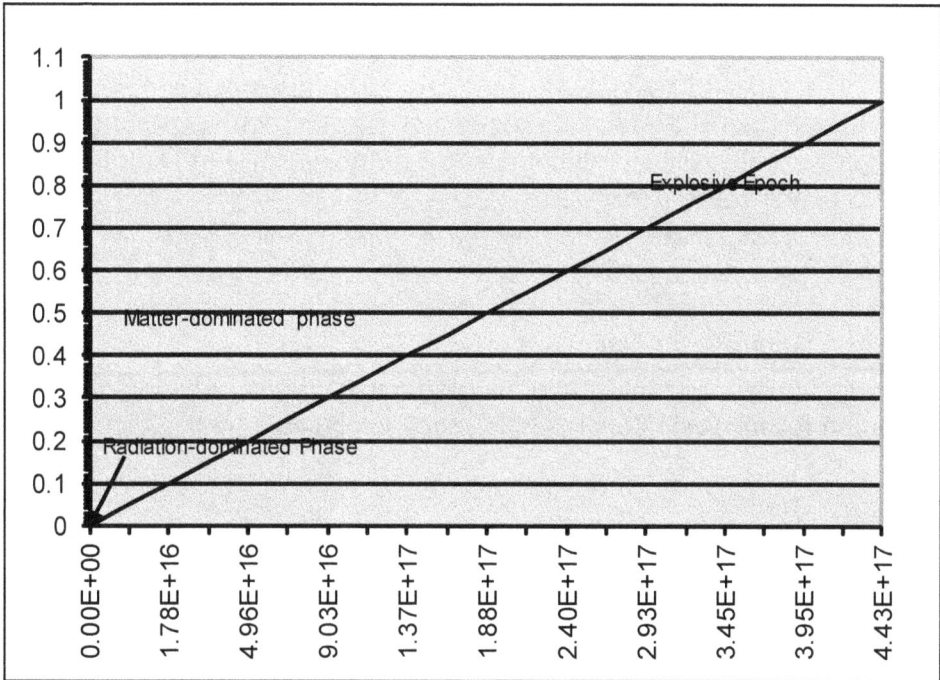

Figure 13.2.1.1. A plot of a(t) generated from eq. 13.2.1.4 through numerical integration.

Epoch	Type	Phases
I	**Explosive Growth**	
II	**Expanding**	**Matter-Dominated** **Radiation-Dominated**

Table 13.2.1.1. Epochs and phases of the universe after the Big Bang Epoch.

The beginning of this epoch is set by eq. 13.2.2.3:

$$t_E = t_{now} + (H_0\Omega_\Lambda^{1/2})^{-1}\ln(.76) = 2.87 \times 10^{17} \text{ s} \qquad (13.2.2.7)$$

where $t_{now} = 4.35 \times 10^{17}$ s giving the time interval of this epoch as 1.47×10^{17} s $= 4.7$ Gyr —much longer than the radiation-dominated era, and still expanding. The transition time t_E is close to the standard hypothesis for the appearance of Dark Energy of at a red-shift $z \sim .5$ or $a(t) = 2/3$. $(1 + z = a^{-1})$

   Thus the Explosive Growth period begins around 2.87×10^{17} s ($t_E = 9.1$ Gyr), extends 4.7 Gyrs to the present $t_{now} = 4.35 \times 10^{17}$ s (13.8 Gyrs).

13.2.3 The Expanding Epoch

The Expanding Epoch includes the matter-dominated and radiation-dominated phases. The solution for the scale factor in this epoch is obtained by neglecting the Ω_Λ term in eq. 13.2.1.4:

$$\int_{a(t')}^{a(t)} da\, a\, H_0^{-1} \left[-\xi a^2 + \Omega_m a + \Omega_\gamma \right]^{-1/2} = \int_{t'}^{t} dt \qquad (13.2.3.1)$$

This equation is easily integrated yielding:

$$-\xi H_0(t - t') = [-\xi a^2(t) + \Omega_m a(t) + \Omega_\gamma]^{1/2} + \Omega_m(2\xi^{1/2})^{-1}\arcsin[(\Omega_m - 2\xi a(t))(\Omega_m^2 + 4\xi\Omega_\gamma)^{-1/2}] - $$
$$- [-\xi a^2(t') + \Omega_m a(t') + \Omega_\gamma]^{1/2} - \Omega_m(2\xi^{1/2})^{-1}\arcsin[(\Omega_m - 2\xi a(t'))(\Omega_m^2 + 4\xi\Omega_\gamma)^{-1/2}]$$
$$(13.2.3.2)$$

Letting $t' = 0$ assuming $a(t' = 0) = 0$ and we find

$$-\xi H_0 t = [-\xi a^2(t) + \Omega_m a(t) + \Omega_\gamma]^{1/2} + \Omega_m(2\xi^{1/2})^{-1}\arcsin[(\Omega_m - 2\xi a(t))(\Omega_m^2 + 4\xi\Omega_\gamma)^{-1/2}] - $$
$$- \Omega_\gamma^{1/2} - \Omega_m(2\xi^{1/2})^{-1}\arcsin[\Omega_m(\Omega_m^2 + 4\xi\Omega_\gamma)^{-1/2}] \qquad (13.2.3.3)$$

Eq. 13.2.3.3 can be substantially simplified. Note the arguments of the arcsines are both near one in value due to the smallness of Ω_γ and ξ. Both arcsines can be approximated using

$$\arcsin(1 - \epsilon) \cong \pi/2 - (2\epsilon)^{\frac{1}{2}} \qquad (13.2.3.4)$$

for small ϵ.

First we approximate the arguments of the arcsines with

$$\arcsin[(\Omega_m - 2\xi a(t))(\Omega_m^2 + 4\xi\Omega_\gamma)^{-\frac{1}{2}}] \cong \arcsin(1 - 2\xi a(t)\Omega_m^{-1} - 2\xi\Omega_\gamma\Omega_m^{-2})$$

and

$$\arcsin[\Omega_m(\Omega_m^2 + 4\xi\Omega_\gamma)^{-\frac{1}{2}}] \cong \arcsin(1 - 2\xi\Omega_\gamma\Omega_m^{-2})$$

Then using eq. 13.2.3.4 we obtain

$$[-\xi a^2(t) + \Omega_m a(t) + \Omega_\gamma]^{\frac{1}{2}} - [\Omega_m a(t) + \Omega_\gamma]^{\frac{1}{2}} \cong -\xi H_0 t \qquad (13.2.3.5)$$

Noting that $\xi a^2(t)$ is much smaller than the other terms in the first square root in eq. 13.2.3.5 we can further approximate that equation further by expanding the square root to obtain:

$$a^2(t)[\Omega_m a(t) + \Omega_\gamma]^{-\frac{1}{2}} \cong 2H_0 t \qquad (13.2.3.6)$$

Eq. 13.2.3.6 embodies the standard matter-dominated and radiation-dominated expressions for the scale factor.

Matter-Dominated Phase

In the case of the matter-dominated phase we have

$$\Omega_m a(t) > \Omega_\gamma$$

and can approximate eq. 13.2.3.6 accordingly

$$a^2(t)[\Omega_m a(t)]^{-\frac{1}{2}} \cong 2H_0 t \qquad (13.2.3.7)$$

or

$$a(t) \cong [2H_0\Omega_m^{\frac{1}{2}}t]^{2/3} \qquad (13.2.3.8)$$

Radiation-Dominated Phase

In the case of the radiation-dominated phase we have

$$\Omega_m a(t) < \Omega_\gamma$$

due to the smallness of the scale factor in that phase We approximate eq. 13.2.3.6 accordingly

$$a^2(t)[\Omega_\gamma]^{-\frac{1}{2}} \cong 2H_0 t \qquad (13.2.3.9)$$

Thus

$$a(t) \cong [2H_0\Omega_\gamma^{\frac{1}{2}}t]^{\frac{1}{2}} \qquad (13.2.3.10)$$

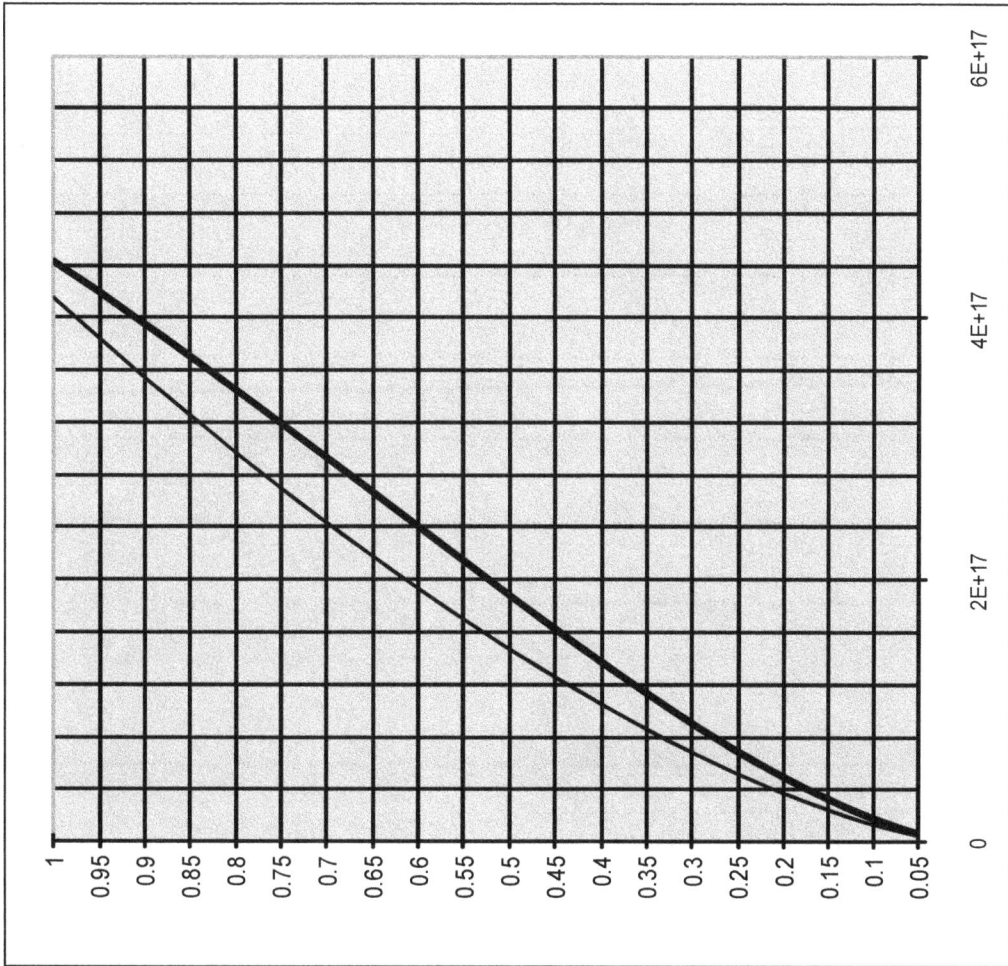

Figure 13.2.3.1. A plot of a(t) (horizontal axis) vs. time in seconds. The thick line is the plot of a(t) obtained by direct numerical integration of eq. 13.2.1.4 including the three density terms and the curvature constant term. The thin line is a plot of a(t) calculated directly from the approximation eq. 13.2.3.6. The approximation becomes increasingly better for small times as t → 0. (Reader: please rotate page 90 degrees clockwise.)

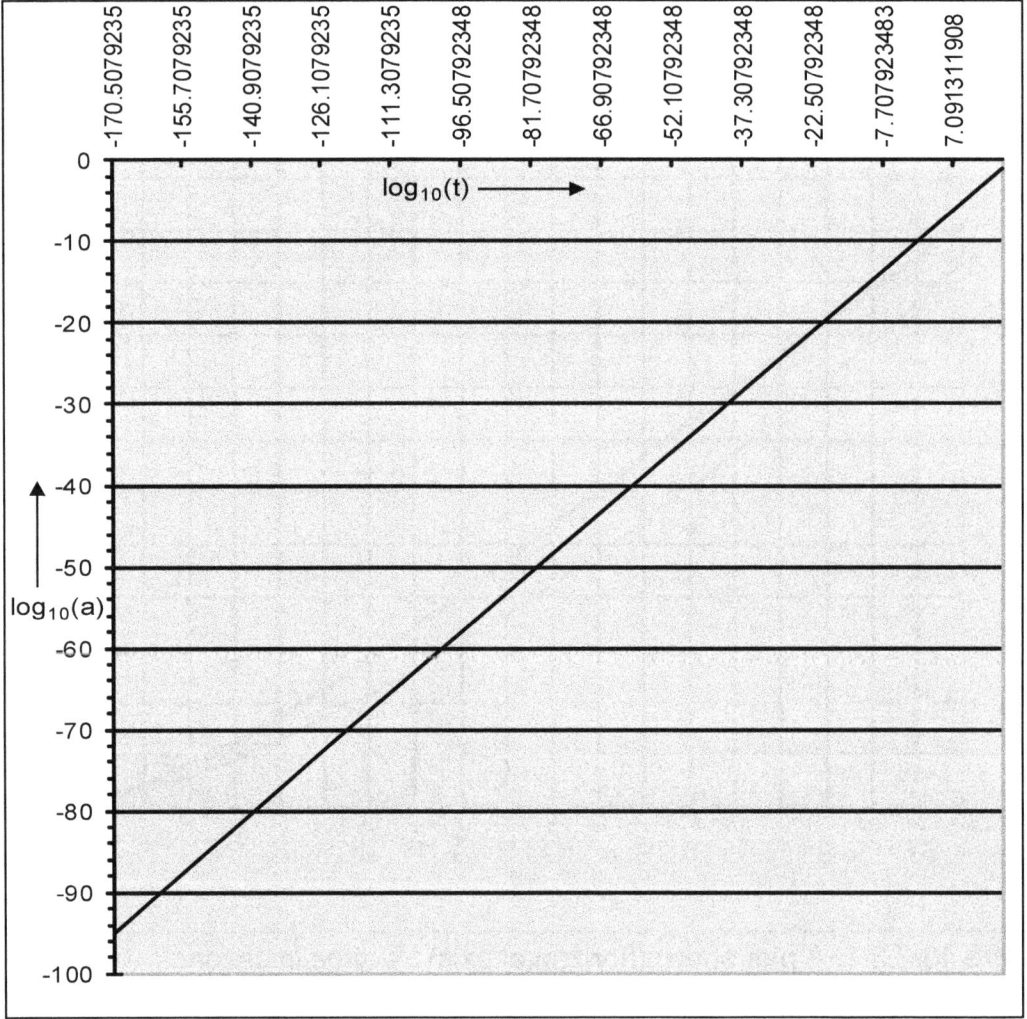

Figure 13.2.3.2. A log-log (base 10) plot of a(t) vs. t (in seconds) for small times calculated from eq. 13.2.3.6.

The crossover point between the radiation-dominated and matter-dominated phase is at

$$\Omega_m a(t_{RM}) = \Omega_\gamma \quad \text{or} \quad a(t_{RM}) = 1.8 \times 10^{-4} \qquad (13.2.3.11)$$

where t_{RM} is the crossover time:

$$t_{RM} = 7 \times 10^{11} \text{ s} = 2.22 \times 10^{-5} \text{ Gyr} \qquad (13.2.3.12)$$

which indicates a very short radiation-dominated phase. Figs. 13.2.3.1 and 13.2.3.2 contain plots of a(t). Fig. 13.2.3.1 shows the approximate implicit equation for a(t) (eq. 13.2.3.6) is quite good over the entire range, and particularly good for small times in the radiation-dominated time frame. Since this region is the region of interest as it connects to the Big Bang Epoch we shall use this approximation, and eq. 13.2.3.10, for a(t) at very small times near t = 0.

13.3 Two-Tier Quantum Big Bang Model at the Beginning of Time

Section 13.2 presented the approximate expressions for the scale factor near t = 0. In this section we will consider numerical estimates for the scale factor, and other quantities of interest, such as the temperature and density near, and at, t = 0 in the Two-Tier Hybrid Big_Bang model whose early time behavior is described in chapter 11.

In estimating quantities we are confronted with an imprecise determination of the needed input parameters because of experimental uncertainties and the impossibility of currently finding certain parameters experimentally in a model independent way. So we will use reasonable estimates for input parameters realizing that they may sometimes be off by up to a few orders of magnitude. Because of the vast differences in value between terms comprising the scale factors we believe a few orders of magnitude is often not a significant issue in determining the relative importance of terms.

The reader may notice slight differences in values due to rounding off numbers to three figures in the text while keeping values to 16 significant digits in the calculations. These differences can have a cumulative effect so we want to emphasize that our goal is order of magnitude accuracy.

The input data items are those of section 13.2 plus:

1. The current temperature of the Cosmic Microwave Background (CMB) radiation = 2.7255 °K.

2. We assume $M_c = M_{Planck} = 1.22 \times 10^{28}$ eV since M_{Planck} is the only large mass intrinsic to the theory of gravitation and thus seemed to be a natural choice.

13.3.1 The CMB Temperature

From the current CMB temperature (T = 2.7255 °K) we find

$$\kappa T = \kappa T_0/a(t_{now}) = \kappa T_0 = 2.35 \times 10^{-4} \text{ eV} \qquad (13.3.1.1)$$

For later use we define

$$\chi = \pi^{3/2}\kappa T_0/(2M_c) \cong 5.3 \times 10^{-32} \qquad (13.3.1.2)$$

13.3.2 The Generalized Robertson-Walker Scale Factor

A_{BB} is related to B_{BB} by eq. 11.86. Our approximation for the B_{BB} scale factor is:[72]

$$B_{BB}(t, y) \cong (1 + M_c^2 y^2)^{-1}\{-i\chi M_c y + k^{-\frac{1}{2}}M_c a(t) + [\chi^2 - 2i\chi y k^{-\frac{1}{2}}M_c^2 a(t) + k^{-1}M_c^2 a^2(t)]^{\frac{1}{2}}\}$$

$$(13.1.5)$$

If we let $y = M_c^{-1}$ then we can find B_{BB} "at the borders of the universe" where it simplifies to:

$$B_{BB}(t) = B_{BB}(t, M_c^{-1}) \cong \{-i\chi + \varpi a(t) + [\chi^2 - 2i\chi\varpi a(t) + \varpi^2 a^2(t)]^{\frac{1}{2}}\}/2 \quad (13.3.2.1)$$

with

$$\varpi = k^{-\frac{1}{2}}M_c = 8.31 \times 10^{60} \qquad (13.3.2.2)$$

If t and a(t) are small,

$$A_{BB} = A_{BB}(t, \check{r}) = M_c^{-1}B_{BB}(t, \check{r}) \cong \beta_0(\check{r}) + \beta_1(\check{r})a(t) + \dots \qquad (13.3.2.3)$$

---

[72] Please note that the physical value of scale factors is the absolute value of the complex scale factors (obtained by use of the Reality group transformations of Blaha (2018e)) appearing here and in the following discussions.

$$\beta_0(\check{r}) = \varkappa(1 - i\check{r})/[M_c(1 + \check{r}^2)] \qquad (13.3.2.4)$$

$$\beta_1(\check{r}) = k^{-\frac{1}{2}}(1 - i\check{r})/(1 + \check{r}^2) \qquad (13.3.2.5)$$

We noted earlier that if we transformed our results back to Robertson-Walker coordinates we would have a modified scale factor a = a_{BBRW} which we can continue to express in terms of $\check{r} = M_c y$.

$$a(t) \rightarrow (1 + M_c^2 y^2)k^{\frac{1}{2}}A_{BB}(t, y)/2 \equiv a_{BBRW}(t, \check{r}) \qquad (8.5.2.4)$$

Thus

$$a_{BBRW}(t, \check{r}) = (1 + M_c^2 y^2)(2k^{-\frac{1}{2}}M_c)^{-1}B_{BB}(t, y) \qquad (13.3.2.6)$$

$$= \frac{1}{2}\{a(t) - i\varkappa\varpi^{-1}\check{r} + [(\varkappa/\varpi)^2 + a^2(t) - 2i(\varkappa/\varpi)\check{r}a(t)]^{\frac{1}{2}}\} \quad (13.3.2.7)$$

by eq. 8.92. If we evaluate a_{BBRW} at $\check{r} = M_c y = 1$ (the maximum value of y), since it determines the "size" of the universe, then

$$a_{BBRW}(t) \equiv a_{BBRW}(t, 1) = \{a(t) - i\gamma + [\gamma^2 + a^2(t) - 2i\gamma a(t)]^{\frac{1}{2}}\}/2 \qquad (13.3.2.8)$$

with the dimensionless constant

$$\gamma = \varkappa/\varpi = 6.38 \times 10^{-93} \qquad (13.3.2.9)$$

This value reminds one of Eddington's famous remark that cosmological quantities often have orders of magnitude that are approximately multiples of 90.
The real and imaginary parts of $a_{BBRW}(t)$ are:

$$\text{Re } a_{BBRW}(t) = a(t)/2 + [Q(t)(1 + \cos \psi(t))/2]^{\frac{1}{2}}/2 \qquad (13.3.2.10)$$

and

$$\text{Im } a_{BBRW}(t) = -\gamma/2 - [Q(t)(1 - \cos \psi(t))/2]^{\frac{1}{2}}/2 \qquad (13.3.2.11)$$

where

$$Q(t) = [(\gamma^2 + a^2(t))^2 + 4\gamma^2 a^2(t)]^{\frac{1}{2}} \qquad (13.3.2.12)$$

and

$$\cos \psi(t) = (\gamma^2 + a^2(t))/Q \qquad (13.3.2.13)$$

There are 2 distinctly different periods specified by the time dependence of a_{BBRW}. The first period corresponds to an initial universe of "slowly" increasing size. The second period is the radiation and matter dominated phases with a fairly rapid increase in a(t). The boundary time t_c between these periods is specified by a continuity condition using:

$$a(t) \cong [2\Omega_\gamma^{1/2} H_0 t]^{1/2} \qquad (13.2.3.10)$$

above t_c and

$$\text{Re } a_{BBRW}(t_c) = a(t_c)/2 + [Q(t_c)(1 + \cos \psi(t_c))/2]^{1/2}/2 = a(t_c)$$

below using eq. 13.2.3.10. Eq. 13.3.2.14 below is the self-consistency condition[73]

$$[2\Omega_\gamma^{1/2} H_0 t]^{1/2} = a(t_c) \cong a_{BBRW}(t_c) = \gamma \qquad (13.3.2.14)$$

giving

$$t_c = 1.26 \times 10^{-165} \text{ s} \cong 10^{-165} \text{ s} \qquad (13.3.2.15)$$

In this period we find

$$\text{Re } a_{BBRW}(0) \cong \gamma/2 = 3.19 \times 10^{-93} \quad (13.3.2.16a)$$

$$\text{Re } a_{BBRW}(t_c) \cong 1.28\gamma = 8.166 \times 10^{-93} \qquad (13.3.2.16b)$$

$$\text{Im } a_{BBRW}(0) = -\gamma/2 = -3.19 \times 10^{-93} \qquad (13.3.2.16c)$$

---

[73] This continuity condition is based on the relation a(t) $a(t) \rightarrow a_{BBRW}(t, \v{r})$ in eq. 13.4.1 a(t) is embedded within $a_{BBRW}(t, \v{r})$. **The continuity condition implies that the Y black body energy becomes the energy Ω_γ.** $_\gamma$

with the radius, and volume, of the universe "slowly" increasing during this period. The nature of this period directly reflects the effects of the blackbody Y quanta. This can be seen from its inverse square dependence on M_c. As $M_c \to \infty$ the constant $\gamma \to 0$ and thus $a_{BBRW} \to 0$, with the universe scaling down to zero size with the attendant standard model catastrophes of infinite density and temperature., The blackbody quanta (Dark Energy) give the universe a short-lived meta-stable initial size thus avoiding catastrophic divergences.

Epoch	Type	Phases	Time Period
I	Explosive Growth	Dark Energy-dominated	$2.87 \times 10^{17}\,\text{s} - 4.35 \times 10^{17}\,\text{s}$
II	Expanding	Matter-dominated	$7 \times 10^{11}\,\text{s} - 2.87 \times 10^{17}\,\text{s}$
		Radiation-dominated	$1.26 \times 10^{-165}\,\text{s} - 7 \times 10^{11}\,\text{s}$
III	Metastable Big Bang	Blackbody Y quanta dominated	$0\,\text{s} - 1.26 \times 10^{-165}\,\text{s}$

Epoch	Type	Phases	Time Interval
I	Explosive Growth	Dark Energy-dominated	4.7 Gyr
II	Expanding	Matter-dominated	9.1 Gyr
		Radiation-dominated	$2.22 \times 10^{-5}\,\text{Gyr}$
III	Metastable Big Bang	Blackbody Y quanta dominated	$1.26 \times 10^{-165}\,\text{sec}$

Table 13.3.2.1 Epochs and phases of the Universe since t = 0.

The period after t_c is dominated by the usual scale factor a(t). This scale factor shows up directly in the real part of a_{BBRW}. Its behavior is:

$t < t_c$ (or a(t) < γ)

$$\text{Re } a_{BBRW}(t) \cong \gamma/2 + a(t)/2 \qquad (13.3.2.17a)$$

$t_c < t < t_{now}$

$$\text{Re } a_{BBRW}(t) \cong a(t) \qquad (13.3.2.17b)$$

The behavior of the imaginary part of a_{BBRW} is also indirectly dominated by a(t) but in a much less dramatic way. The gradual growth of the imaginary part of a_{BBRW} is more or less indicated by the following three values:

$$\text{Im } a_{BBRW}(0) \cong -\gamma/2 = -3.19 \times 10^{-93} \qquad (13.3.2.16c)$$
$$\text{Im } a_{BBRW}(t_c) \cong -0.822\gamma = -5.24 \times 10^{-93} \qquad (13.3.2.18a)$$
$$\lim_{t \to \infty} \text{Im } a_{BBRW}(t) = -\gamma = -6.38 \times 10^{-93} \qquad (13.3.2.18b)$$

and also the behavior

a(t) $\ll$ γ (or t < t_c)

$$\text{Im } a_{BBRW}(t) \cong -\gamma/2 - a(t)/2 \qquad (13.3.2.18c)$$

$\gamma \ll$ a(t) (or t_c < t)

$$\text{Im } a_{BBRW}(t) \cong -\gamma + \mathcal{O}([\gamma/a(t)]^2) \cong -\gamma \qquad (13.3.2.18d)$$

Both the real and imaginary parts of a_{BBRW} roughly double in the time period [0, t_c] and thereafter we see a gradual increase of Im a_{BBRW} in absolute value from $-.5\gamma$ to $-\gamma$ over the lifetime of the universe.

After t_c the real part grows dramatically (eq. 13.3.2.17) while the imaginary part remains minute. The details of the interpretation of the behavior of the scale factor a_{BBRW} in the Big Bang Epoch [0, t_c] will be explored in chapter 15. It suffices, for now, to say the universe in the period before t_c is in a meta-stable state of "slowly" growing size due to the dynamics of the blackbody Y quanta. At t_c the epoch of the expanding universe, as we know it, begins!

In differentiating between the real and imaginary parts of a_{BBRW} we must realize that the Reality group combines them into a single real quantity when scaling coordinates. Happily the small size of the imaginary part it can be neglected in most situations.

Before proceeding to describe physically interesting features of the early universe we note the radial dependence of the scale factor $a_{BBRW}(t, \check{r})$ in general:

$$a_{BBRW}(t, \check{r}) = \{-i\gamma\check{r} + a(t) + [\gamma^2 - 2i\gamma\check{r}a(t) + a^2(t)]^{\frac{1}{2}}\}/2 \quad (13.3.2.19)$$

$$\text{Re } a_{BBRW}(t, \check{r}) = a(t)/2 + [R(t, \check{r})(1 + \cos\psi(t, \check{r})/2]^{\frac{1}{2}}/2 \quad (13.3.2.20)$$

$$\text{Im } a_{BBRW}(t, \check{r}) = -\gamma\check{r}/2 - [R(t, \check{r})(1 - \cos\psi(t, \check{r}))/2]^{\frac{1}{2}}/2 \quad (13.3.2.21)$$

where

$$R(t, \check{r}) = [(\gamma^2 + a^2(t))^2 + 4(\gamma\check{r}a(t))^2]^{\frac{1}{2}} \quad (13.3.2.22)$$

and

$$\cos\psi(t, \check{r}) = (\gamma^2 + a^2(t))/R(t, \check{r}) \quad (13.3.2.23)$$

We find that the scale factor $a_{BBRW}(t, \check{r})$ is approximated in various time periods to well within an order of magnitude by

$0 \le t < t_c$

$$\text{Re } a_{BBRW}(t, \check{r}) \cong \gamma/2 + a(t)/2 \quad (13.3.2.24)$$

$$\text{Im } a_{BBRW}(t, \check{r}) \cong -\gamma\check{r}/2 - a(t)\check{r}/2 \quad (13.3.2.25)$$

$t_c < t$

$$\text{Re } a_{BBRW}(t, \check{r}) \cong a(t)\{1 + (1 + \check{r}^2)\gamma^2/4\} \cong a(t) \quad (13.3.2.26)$$

$$\text{Im } a_{BBRW}(t, \check{r}) \cong -\gamma\check{r}/2 - [\check{r}^2 + \gamma^2/(4a^2(t))]^{\frac{1}{2}}\gamma/2 \quad (13.3.2.27)$$

$t \to \infty$

$$\text{Re } a_{BBRW}(t, \check{r}) \cong a(t)\{1 + (1 + \check{r}^2)\gamma^2/4\} \cong a(t) \quad (13.3.2.28)$$

$$\text{Im } a_{BBRW}(t, \check{r}) \cong -\gamma\check{r} = -8.166 \times 10^{-93}\check{r} \quad (13.3.2.29)$$

13.3.3 Temperature of the Early Universe in the Generalized Robertson-Walker Metric

We will now examine the temperature of the early universe based on the results of the preceding section. For t near t = 0, we note κT depends on the blackbody scale factor of the generalized Robertson-Walker metric:[74]

$$\kappa T = \kappa T_0 / B_{BB}(t, y) \tag{13.3.3.1}$$

and

$$B_{BB}(t, y) = 2k^{-\frac{1}{2}}M_c(1 + M_c^2 y^2)^{-1} a_{BBRW}(t, \check{r}) \tag{13.3.3.2}$$

by eq. 13.3.2.6 using the variable

$$\check{r} \equiv M_c y \tag{13.3.3.3}$$

for convenience. Therefore from eqs. 13.3.2.24 – 13.3.2.27 we see

$0 \le t < t_c$

$$\begin{aligned}
\kappa T_< \equiv \kappa T(0 \le t < t_c) &\cong 2\kappa T_0(1 + i\check{r})/x = 4\pi^{-3/2}\, M_c(1 + i\check{r}) \\
&= 2.72 \times 10^{29}(1 + i\check{r})\ eV \\
&= 3.155 \times 10^{33}(1 + i\check{r})\,^\circ K
\end{aligned} \tag{13.3.3.4}$$

$t_c < t$

$$\begin{aligned}
\kappa T_> \equiv \kappa T(t_c < t) \\
\cong \kappa T_0(a(t) + i\gamma\check{r})/a^2(t) = \kappa T_0/a(t)
\end{aligned} \tag{13.3.3.5}$$

We note that the temperatures calculated in this section are in Extended Robertson-Walker coordinates (eq. 8.2.4.5) and not in Robertson-Walker coordinates.

---

[74] Here again we remind the reader that complex values for radii, temperature and so on are transformed by the Reality group to real values – their absolute values.

Note also that the physical temperature is the absolute value of the complex temperature due to the use of the Reality group.

13.3.4 Consistency of Y_{BB} Approximation with the Temperature near t = 0

We now address the question of whether the values found for κT are consistent with the approximations made in chapter 11 in order to obtain an expression for Y_{BB}:

$$\cos(\omega\kappa Tt) \approx 1 \qquad (8.66)$$

$$J_1(\omega\kappa Ty) \approx \omega\kappa Ty/2 \qquad (8.67)$$

The first approximation is valid if $\kappa Tt \sim 0$ since the Planck distribution factor makes the largest contribution to the integral come from small ω. The values of $\kappa T_<$ and $\kappa T_>$ show that for larger times κTt is very small:

$$\kappa T_< t_c \leq 3.42 \times 10^{-136}$$

$$\kappa T_> t \leq \kappa T_0 t/a(t) \qquad (13.3.4.1)$$

The second approximation is valid if $\kappa Ty \ll 1$ since the Planck distribution factor again makes the largest contribution to the integral come from small ω. We find the maximum values of κTy in the two time periods from eqns. 13.3.3.4 and 13.3.3.5 to be

$0 \leq t < t_c$

$$MAX(|\omega\kappa T_< y|) = |\omega\kappa T_< M_c^{-1}| = 22.27\omega \qquad (13.3.4.2)$$

which is small since the dominant part of the integration comes from small ω; and

$t_c < t$

$$MAX(|\omega\kappa T_> y|) = |\omega\kappa T_>(t_c)M_c^{-1}| = 2.35 \times 10^{-4}\omega/a(t) \qquad (13.3.4.3)$$

which is also small. Thus the Bessel function power series expansion is well approximated by its first term:

$$J_1(\omega\kappa Ty) \approx (\omega\kappa Ty/2) \qquad (13.3.4.4)$$

We conclude our approximate calculation of A(t, y) is valid for all time and for the complete range of y values: $0 \leq y \leq M_c^{-1}$.

13.3.5 Plots of the Scale Factor from t = 0 to the Present

We will create several plots of the scale factor from t = 0 to the present time t = 4.35 × 10^{17} s for the maximum value of y = M_c^{-1}, which corresponds to the maximum Robertson-Walker radius coordinate value r = $k^{-\frac{1}{2}}$. The approximation that we have developed for B_{BB} has been justified for times between the Big Bang and the present.

In Fig. 13.3.5.1 we show a log – log plot of the real and imaginary parts of $a_{BBRW}(t)$ vs. t using base 10 logarithms for t ∈ $[10^{-200}, 10^{20}]$ seconds. In Fig. 13.3.5.2 we plot the real and imaginary parts of a_{BBRW} vs. time from t = 0 to t = 1.2 × 10^{-246} s. In Fig. 13.3.5.3 we plot the real part of a_{BBRW}, and the Robertson-Walker scale factor a(t), vs. time in seconds in the period around 10^{-167} s. It shows the rapidity of the transition of Re a_{BBRW} from slowly rising to rapidly rising with a(t).

13.4 The Interpretation of the Complex Scale Factor

The interpretation of the complex scale factor $a_{BBRW}(t, \check{r})$ hinges on its role in the expression for the proper interval. The expression for the proper interval in the Robertson-Walker metric is:

$$d\tau^2 = dt^2 - R^2(t)[dr^2/(1 - kr^2) + r^2(d\theta^2 + \sin^2\theta d\varphi^2)]$$

It was extended to

$$d\tau^2 = dt^2 - A^2(t, \check{r})[d\check{r}^2 + \check{r}^2(d\theta^2 + \sin^2\theta \, d\varphi^2)] \qquad (10.4.9)$$

where an identification was made between the Robertson-Walker scale factor R(t) ≡ $a(t)k^{-\frac{1}{2}}$ and $a_{BBRW}(t, \check{r})$ through the following chain of equalities and correspondences

$$a(t) = a(t)b_0(r) = A(t, \check{r})/b(\check{r}) \rightarrow A_{BB}(t, \check{r})(1 + \check{r}^2)k^{\frac{1}{2}}/2 = a_{BBRW}(t, \check{r}) \qquad (13.4.1)$$

where the $\rightarrow$ relates the classical expressions on the left with the expressions on the right that embody quantum corrections due to Y-quanta blackbody radiation. Eq. 13.4.1 is based on eqns. 10.4.6, 10.4.7, 10.4.9, 10.4.11, 10.4.12, and 10.6.4a. Combining the equations behind eq. 13.4.1 we find

$$d\tau^2 = dt^2 - a_{BBRW}(t, \check{r}(r))^2[dr^2/(1 - kr^2) + r^2(d\theta^2 + \sin^2\theta d\varphi^2)] \qquad (13.4.2)$$

where the relation between $\check{r}$ and r, $\check{r}(r)$, is specified by eq. 10.4.12. We call the metric in eq. 13.4.2 the Extended Robertson-Walker metric. It is equivalent to eq. 10.4.9, differing only in the definition of the radial coordinate.

Having determined the role of the complex scale factor in the metric tensor we can now physically interpret it by applying a Reality group transformation that effectively multiplies $a_{BBRW}(t, \check{r}(r))$ by a phase making it real and equal to its absolute value $|a_{BBRW}(t, \check{r}(r))|$.

We can simplify the physical interpretation without loss of generality by considering the radial coordinate of eq. 10.4.9 to be

$$r_{GRW} = A_{BB}\check{r} \qquad (13.4.3)$$

$$= r_r + ir_i \qquad (13.4.4)$$

where

$$A_{BB}(t, y) = 2a_{BBRW}(t, \check{r})/[(1 + M_c^2y^2)k^{\frac{1}{2}}] \qquad (8.5.2.4)$$

Thus

$$r_{GRW} = 2a_{BBRW}(t, \check{r})\check{r}/[(1 + \check{r}^2)k^{\frac{1}{2}}] \qquad (13.4.5)$$

Applying a Reality group transformation we obtain the physically measurable, real-valued radius

$$r_{GRWphysical} = 2|a_{BBRW}(t, \check{r})\check{r}/[(1 + \check{r}^2)k^{\frac{1}{2}}]| \qquad (13.4.5a)$$

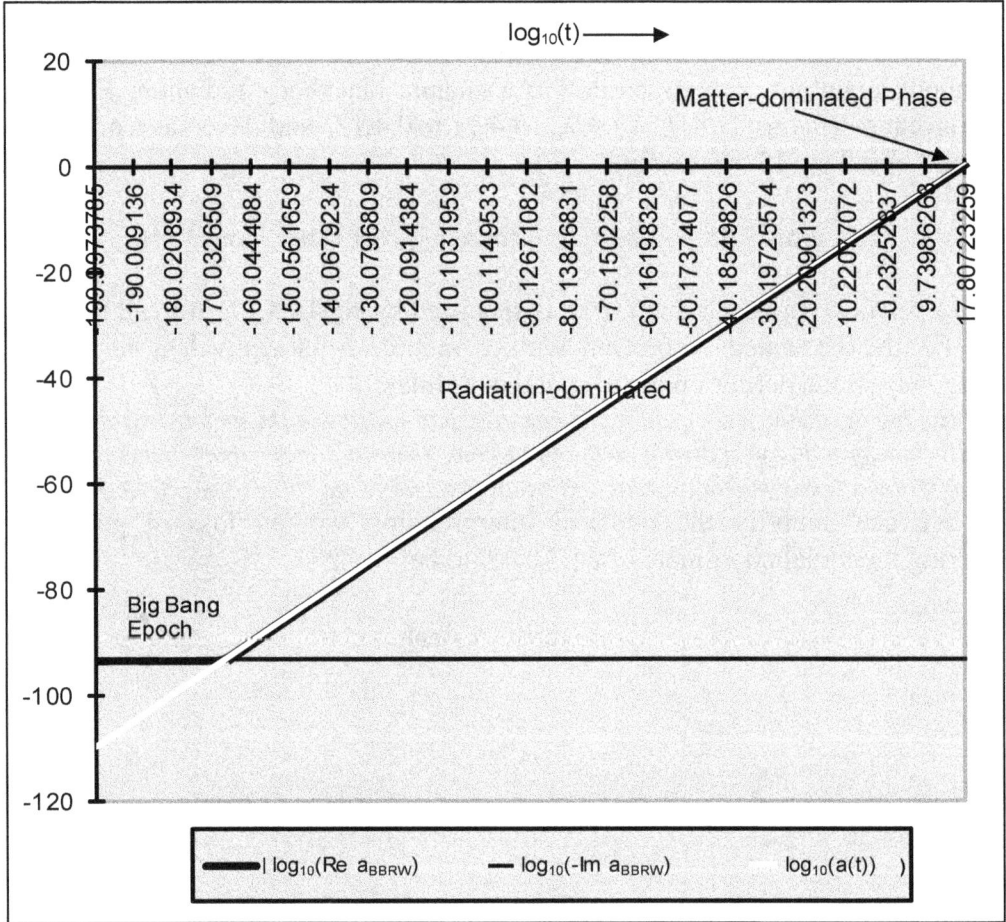

Figure 13.3.5.1. A log-log (base 10) plot of the real and imaginary parts of a_{BBRW} and the Robertson-Walker scale factor a(t) versus the log (base 10) of time in seconds. Note the imaginary part of a_{BBRW} is very slowly growing. The real part of a_{BBRW} is growing slowly until $t_c \approx 10^{-165}$ s and thereafter equals a(t) to good approximation.

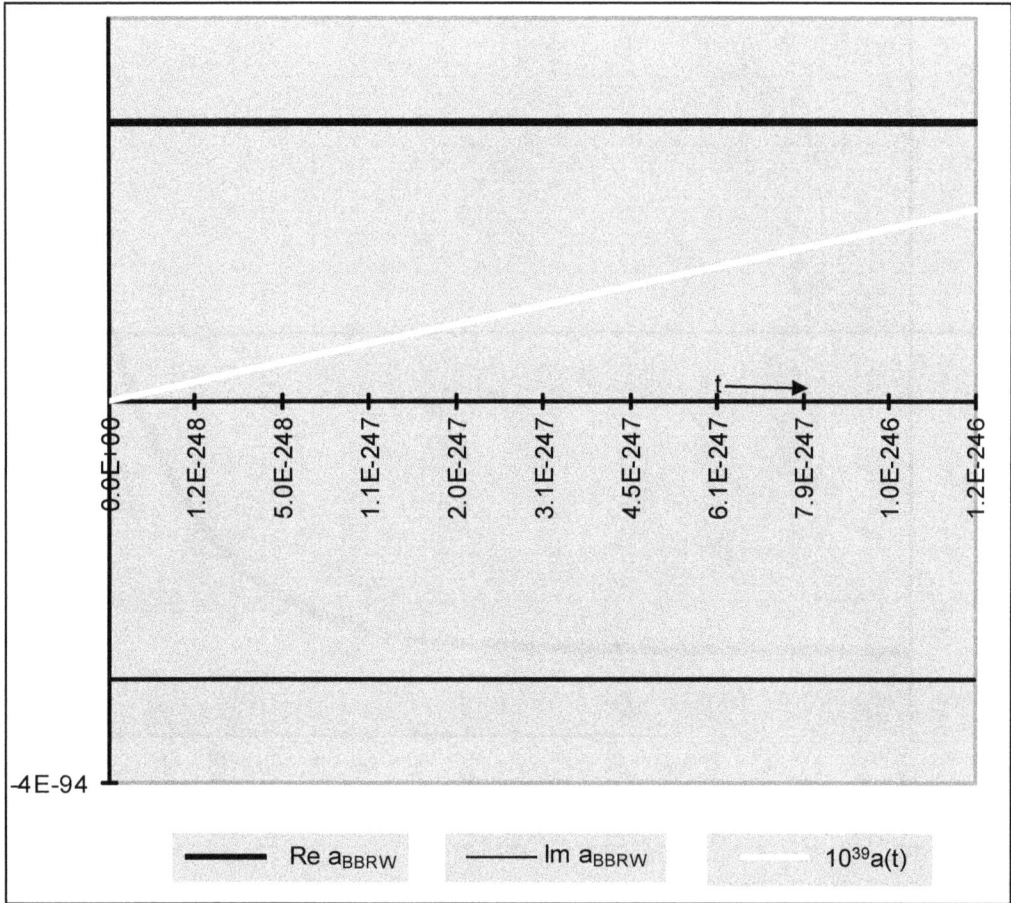

Figure 13.3.5.2. A plot of the real and imaginary parts of a_{BBRW}, and $10^{39} \times a(t)$, versus time from t = 0 to t = 1.2×10^{-246} s. Note they are slowly varying and well behaved in the neighborhood of t = 0 with only a(t) having the value of zero. "E" indicates a power of ten (for example: 2.0E-248 = 2.0×10^{-248} s).

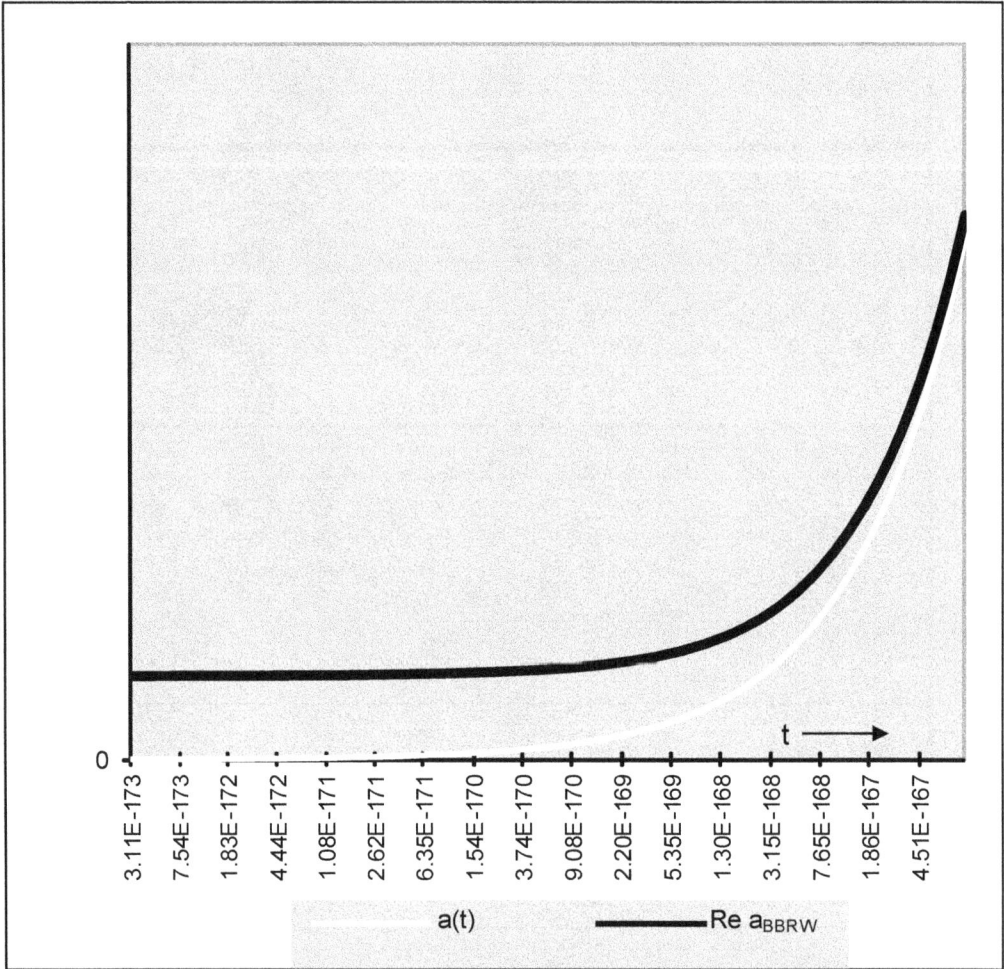

Figure 13.3.5.3. A plot of the real part of a_{BBRW} and a(t) vs. time in seconds around the time 10^{-165} s. Note Re a_{BBRW} quickly changes from slowly growing to growing like a(t).

13.5 Time Evolution of the Hubble Rate

The Hubble rate is one of the linchpins of modern cosmology. It is determined by Einstein's equation eq. 13.2.1.1 when written in the form:

$$H(t) = \dot{a}/a = [H_0^2(\Omega_\gamma/a^4(t) + \Omega_m/a^3(t) + \Omega_\Lambda) - k/a^2(t)]^{\frac{1}{2}} \tag{13.5.1}$$

At small times, in the radiation-dominated phase, eq. 13.5.1 can be approximated by

$$H(t) \cong H_0\Omega_\gamma^{\frac{1}{2}}/a^2(t) \tag{13.5.2}$$

If we define a Hubble rate H(t) using $a_{BBRW}(t, \check{r})$ then

$$H_{BBRW}(t, \check{r}) = |\dot{a}_{BBRW}(t, \check{r})/a_{BBRW}(t, \check{r})| \equiv |\dot{A}_{BBRW}(t, \check{r})/A_{BBRW}(t, \check{r})|$$
$$= |[H_0^2(\Omega_\gamma/a_{BBRW}^4(t, \check{r}) + \Omega_m/a_{BBRW}^3(t, \check{r}) + \Omega_\Lambda) - k/a_{BBRW}^2(t, \check{r})]^{\frac{1}{2}}| \tag{13.4.3}$$

$H_{BBRW}(t, \check{r})$ is the same as H(t) until we reach the first instants of the universe, which we have called the Big Bang Epoch., Then we find

$0 \leq t < t_c$

$$H_{BBRW}(t, \check{r}) \cong H_0\Omega_\gamma^{\frac{1}{2}}/|a_{BBRW}(t, \check{r})|^2 \tag{13.4.4}$$

where

$$\text{Re } a_{BBRW}(t, \check{r}) \cong \gamma/2 + a(t)/2 \tag{13.3.2.24}$$

and

$$\text{Im } a_{BBRW}(t, \check{r}) \cong -\gamma\check{r}/2 - a(t)\check{r}/2 \tag{13.3.2.25}$$

Substituting in eq. 13.4.4 we find

$$H_{BBRW}(t, \check{r}) \cong 4H_0\Omega_\gamma^{\frac{1}{2}}|[1 - \check{r}^2 + 2i\check{r}]/[(\gamma + a(t))^2(1 + \check{r}^2)^2]| \tag{13.4.5}$$

$|H_{BBRW}(t, \check{r})|$ is a real physical number in this range. Thus space has a Hubble rate that is both space and time dependent in the Big Bang Epoch.

13.6 Hubble Constant in the Big Bang Metastate at t = 0 and t = t_c

13.6.1 At Time t = 0

At t = 0 we find $H_{BBRW}(0, \check{r})$ is finite unlike the radiation-dominated Hubble rate (eq. 13.5.2):

$$H_{BBRW}(0, \check{r}) \cong 4H_0\Omega_{,\gamma}^{1/2}|[1 - \check{r}^2 + 2i\check{r}]/[\gamma^2(1 + \check{r}^2)^2]| \qquad (13.4.6)$$

At the "edge" of the universe the Hubble rate is

$$H_{BBRW}(0, 1) \cong 2H_0\Omega_{,\gamma}^{1/2}/\gamma^2 = 5.8 \times 10^{198} \text{ s}^{-1} \qquad (13.4.7)$$
$$= 1.79 \times 10^{218} \text{ km s}^{-1} \text{ Mpc}^{-1}$$

(compared to the current Hubble constant of 100h km s^{-1} Mpc^{-1}), and is solely due to the imaginary part of $4H_0\Omega_{,\gamma}^{1/2}[1 - \check{r}^2 + 2i\check{r}]/[(\gamma + a(t))^2(1 + \check{r}^2)^2]$ since the real part is zero. At $\check{r} = 0$,

$$H_{BBRW}(0, 0) \cong 4H_0\Omega_{,\gamma}^{1/2}1/\gamma^2 = 2H_{BBRW}(0, 1) \qquad (13.4.8)$$

is more rapidly expanding.

Thus we consistently avoid the divergences that appear at t = 0 in the Standard Cosmological Model. Its radiation-dominated phase's Hubble rate is $(2t)^{-1}$ which diverges at t = 0.

13.6.2 At Time t = t_c

At the "edge" of the universe the enormous Hubble rate at the exit time $t_c = 1.26 \times 10^{-165}$ s from the Big Bang metastate is[75]

$$H_{BBRW}(t_c, 1) \cong 4H_0\Omega_{,\gamma}^{1/2}|i|/[2(\gamma + a(t_c))^2]| \qquad (13.4.9)$$
$$= 2 \ H_0\Omega_{,\gamma}^{1/2}/(\gamma + \gamma^2) \cong 2 \ H_0\Omega_{,\gamma}^{1/2}/\gamma = 3.7 \times 10^{106} \text{ s}^{-1}$$
$$= 1.14 \times 10^{126} \text{ km s}^{-1} \text{ Mpc}^{-1}$$

---

[75] Eq. 13.3.2.15.

compared to the current Hubble constant of 100h km s^{-1} Mpc^{-1}.
At ř = 0, the "center" of the Big Bang has

$$H_{BBRW}(t_c, 0) \cong H_0\Omega_{\gamma}^{\frac{1}{2}}/\gamma^2 = 2.9 \times 10^{198} \text{ s}^{-1} \qquad (13.4.10)$$
$$= 8.95 \times 10^{217} \text{ km s}^{-1} \text{ Mpc}^{-1}$$

using eq. 13.3.2.14. The "center" *is more rapidly expanding by a factor of the order of 10^{93}, which is suggestive of a developing much less dense region.*

13.6.3 Hubble Constant Decrease in the Metastable Big Bang State

In the metastable state lasting 10^{-165} s, at the "edge" of the expanding universe, the Hubble constant has shrunk enormously from 5.8×10^{198} s^{-1} to 3.7×10^{106} s^{-1} — a decrease of a factor of 10^{92} — surfacing again Eddington's 90 orders of magnitude conjecture. The origin of the 90 orders of magnitude appears to be in a constant that combines standard constants appearing in our Two-Tier Quantum Gravity based on black body, the Big Bang expansion and the current CMB temperature. The Beginning is thereby directly related to the Present:

$$\gamma = x/\varpi = 6.38 \times 10^{-93} = \pi^{3/2} k^{\frac{1}{2}}\kappa T_0/(2M_c^2) \qquad (13.4.11)$$

Eq. 13.4.11 uses eqs. 13.3.1.2 and 13.3.2.2, and the CMB temperature in.eq. 13.3.1.1.

13.6.4 Creation of a Void in the Metastable State due to a Radially Varying Hubble Constant

Sections 13.6.1 and 13.6.2 show that at t = 0 the Hubble "constant" is a factor of 2 greater at ř = 0 than at the edge of the universe (ř = 1), and at t = t_c (the end of the metastable state life) the Hubble "constant" is a factor of the order of 10^{93} greater at ř = 0 than at the edge of the universe (ř = 1). The vast disparity in Hubble "constant" values at the center compared to the edge of the universe suggests the center will rapidly go to lower density creating a *void. Thus one can expect a "Hubble bubble" generated during the metastable state.*

Today we see numerous voids in the universe.[76] The largest is the spherical KBC supervoid (containing the Milky Way) with a 2 billion light year diameter. Another supervoid is the "Giant void" (Canes Venatici) with a diameter of 1.3 light years. There is also the WMAP "cold spot" void with a diameter of 120 Megaparsecs.

The Hubble Constant within this void is larger due to the attraction of mass-energy external to the void just as we see above when comparing the center to the edge of the Metastable state.

The metastate "void" appears to be a precursor to the later voids seen now.

13.7 Energy Density in the Big Bang Metastable State

An estimate of the density of energy in the Big Bang metastable state can be obtained as follows. At small times, in the radiation-dominated phase, eq. 13.5.1 can be approximated by

$$a'/a = H(t) \cong H_0 \Omega_{\gamma}^{1/2}/a^2(t) \cong H_{BBRW}(t, \check{r}) = H_0 \Omega_{\gamma}^{1/2}/a_{BBRW}^2(t) \qquad (13.5.2)$$

from eq. 13.4.5. The "Big Bang" regime Einstein Equation is

$$a'_{BBRW}{}^2 - 8\pi G \rho_{BBRW} a_{BBRW}{}^2/3 = -k \qquad (13.7.1)$$

implying

$$\rho_{BBRW} = (3/8\pi G)[(a'/a) + k]$$
$$= (3/8\pi G)[H_{BBRW}(t, \check{r}) + k] \qquad (13.7.2)$$

Since

$$H_{BBRW}(t, \check{r}) \cong H_0 \Omega_{\gamma}^{1/2}/|a_{BBRW}(t, \check{r})|^2 \qquad (13.4.4)$$

and

$$\text{Re } a_{BBRW}(t, \check{r}) \cong \gamma/2 + a(t)/2 \qquad (13.3.2.24)$$
$$\text{Im } a_{BBRW}(t, \check{r}) \cong -\gamma\check{r}/2 - a(t)\check{r}/2 \qquad (13.3.2.25)$$

we find

$$\rho_{BBRW}(t = 0, \check{r} = 0) \cong (3/8\pi G)[4H_0 \Omega_{\gamma}^{1/2}/\gamma^2 + k] \cong (3/2\pi\gamma^2 G)H_0 \Omega_{\gamma}^{1/2} \qquad (13.7.3)$$
$$\cong 2.9 \times 10^{194} \text{ gm/cm}^3 = 1.63 \times 10^{218} \text{ GeV/cm}^3$$

---

[76] Zehavi *et al*, Astrophysical Journal **503**, 483 (1998); There are differing results on the Hubble Bubble question such as in Moss *et al*, Phys. Rev. **D83**, 103515 (2011) and references therein.

$$\rho_{\text{BBRW}} \, (t = 0, \, \check{r}=1) \cong (3/8\pi G)[2H_0\Omega_\gamma^{1/2}/\gamma^2 + k] \cong (3/4\pi\gamma^2 G)H_0\Omega_\gamma^{1/2} \tag{13.7.4}$$
$$\cong 1.45 \times 10^{194} \,\, \text{gm/cm}^3 = \, 8.16 \times 10^{217} \,\, \text{GeV/cm}^3$$

At $t = t_c$:
$$\rho_{\text{BBRW}} \, (t = t_c, \, \check{r} = 0) \cong 3/(8\pi G)[H_0\Omega_\gamma^{1/2}/\gamma^2 + k] \cong 3/(8\pi\gamma^2 G)H_0\Omega_\gamma^{1/2} \tag{13.7.5}$$
$$\cong 3.7.26 \times 10^{193} \,\, \text{gm/cm}^3 = \, 4.08 \times 10^{217} \,\, \text{GeV/cm}^3$$

$$\rho_{\text{BBRW}} \, (t = t_c, \, \check{r}=1) \cong 3/(8\pi G)[H_0\Omega_\gamma^{1/2}/(2\gamma^2) + k] \cong 3/(16\pi\gamma^2 G)H_0\Omega_\gamma^{1/2} \tag{13.7.6}$$
$$\cong 3.63 \times 10^{193} \,\, \text{gm/cm}^3 = \, 2.04 \times 10^{217} \,\, \text{GeV/cm}^3$$

14. From t = t_T to the Present: Scale Factor and Hubble Constant

Many years ago Hoyle and Narlikar, and others,[77] proposed a steady state theory of the evolution of the universe in which mass-energy appeared uniformly throughout the universe. They had no convincing explanation for the origin of the influx of mass-energy. In this book we provide a detailed explanation for the source of the influx: the Megaverse. The surface tension force (gravitational) of our universe essentially absorbs mass-energy from the surrounding Megaverse.

In this chapter we will consider the time period from t = t_T = 380,000 to the present during which the influx occurs.[78]

14.1 Influx Epoch Mass-Energy Density and Scale Factor Calculation

Our first concern is to determine $\rho_{tot}(t)$ during the influx time period. We must calculate the total mass-energy density as a function of time including the accumulated total mass-energy influx from t = t_T.

Earlier we found that the derivative of $\rho_{tot}(t)$ (eq. 9.6) is

$$d\rho_{tot}(t)/dt = \rho_{crit}d\Omega_H(t)/dt + C_M k^{\frac{1}{2}}\rho_{tot}(t)/a(t) \qquad (9.9)$$

using $R = k^{-\frac{1}{2}}a(t)$ and

$$C_M = C_T G\rho_{Mega} \qquad (9.8)$$

$$\Omega_H(t) = \Omega_{\gamma}(t) + \Omega_{\Lambda}(t) + \Omega_m(t) + \Omega_d(t) \qquad (9.10)$$

---

[77] H. Bondi and T. Gold, M.N.R.A.S. **108**, 252 (1948); F. Hoyle, M.N.R.A.S. **108**, 372 (1948); F. Hoyle and J. V. Narlikar, Proc. R. Soc. London **A290**, 162 (1966) and references therein.
[78] We selected this time for easy comparison to the known Hubble Constant data.

The solution of eq. 9.9 is

$$\rho_{tot}(t) = \rho_{crit} e^{F(t)} \int_{t_T}^{t} dt'' \, e^{-F(t'')} \, d\Omega_H(t'')/dt'' \qquad (14.1)$$

where

$$F(t) = C_M k^{1/2} \int_{t_T}^{t} dt'/a(t') \qquad (14.2)$$

The above equations for ρ_{to} together with the Einstein equation;

$$\dot{a}^2 - 8\pi G \rho_{tot} a^2/3 = -k \qquad (13.1.2)$$

can be solved simultaneously for the Influx mass-energy density $\Omega_T(t)$, the total density $\rho_{tot}(t)$ as a function of time, and the scale factor $a(t)$ given the values of c_M, T_c, and $\Omega_H(t)$.

14.2 Decomposition of the Mass-Energy Density ρ_{tot} Equation

Eq. 9.9 can be separated into four equations, one equation for each of $\Omega_\gamma(t)$, $\Omega_\Lambda(t)$, $\Omega_m(t)$, $\Omega_d(t)$ and $\Omega_T(t)$. Letting $\Omega_i(t)$ represent each of these densities we find the identities

$$d\Omega_i(t)/dt = d\Omega_i(t)/dt \qquad (14.3)$$

for the first four energy densities of the separations using $\rho_i(t) = \rho_{crit}\Omega_i(t)$, and

$$d\Omega_T(t)/dt = C_M k^{1/2}[\Omega_H(t) + \Omega_T(t)] /a(t) \qquad (14.4)$$

for $\Omega_T(t)$. Eq.14.3 indicates that the energy influx does not change the dynamical equations for $\Omega_H(t)$ but does determine $\Omega_T(t)$.

The solution to eq. 14.4 is

$$\Omega_T(t) = C_M k^{\frac{1}{2}} e^{F(t)} \int_{t_T}^{t} dt''\, e^{-F(t'')} \Omega_H(t'')/a(t'') \tag{14.5}$$

14.2 Megaverse Mass-Energy Densities

The constant Megaverse mass-energy densities will be assumed to be in the same proportions as in our universe at the present time $t = t_{now}$ with

$$\rho_{Mega} = \rho_{Mega0}(\Omega_{\gamma M} + \Omega_{\Lambda M} + \Omega_{mM} + \Omega_{dM})$$

where ρ_{Mega0} is a Megaverse constant "critical density" and[79] the Megaverse energy densities are

$$\begin{aligned}
\Omega_{\gamma M} &= 5.38 \times 10^{-5} \\
\Omega_{\Lambda M} &= 0.692 \\
\Omega_{mM} &= 0.046 \\
\Omega_{dM} &= 0.268
\end{aligned} \tag{14.6}$$

as currently in our universe.

---

[79] M. Tanabashi *et al op. cit.*

15. The Big Bang Epoch

No great thing is created suddenly.
Discourses - *Epictetus*

13.1 The $t = 0$ Big Bang Scenario

The Two-Tier cosmological theory that we have developed in preceding chapters differs dramatically from the Standard Cosmological Model in the Big Bang Epoch and yet smoothly melds into the Standard Cosmological Model in the Expanding Universe and Exploding Universe epochs in chapter 13. Chapter 14 revises the picture significantly when an influx of Megaverse matter is introduced.

The universe has a finite size, temperature and density at the point of the Big Bang which we define to be the time t = 0. The universe grows slowly for a period of time (until roughly 10^{-165} s) that we call the Big Bang metastable state. In this time period we see a very hot, very dense, macroscopic conglomeration of Y quanta, radiation and elementary particles that coexist with each other with non-singular interactions. These particles are not localized. Each particle can be said to occupy the entire universe since the size of the universe is infinitesimal compared to any particle's Compton radius (if it had one). The universe has an almost classical Robertson-Walker type of metric. In particular, it has an Extended Robertson-Walker metric with quantum effects due to a classical Y-quanta blackbody radiation field (the inflatons?) that is both the source of the metastability of the universe at t = 0 and the source of infinitesimal imaginary spatial dimensions that are comparable in extent with the size of the real spatial dimensions of the universe during the Big Bang Epoch.

As the universe expands due to the Y quanta energy it appears that enormous amounts of gravitational energy are also released since the Two-Tier gravitational potential is zero at r = 0 and has a minimum around r ≈ 10^{-33} cm. Thus we view the universe in the Big Bang Epoch as a "slowly" expanding metastable state. (This state is comparable to the metastable false vacuum state in inflation theories – but no scalar bosons are needed – the Y quanta play that role. The combination of gravitation and an effectively classical Y field serve to generate the metastable initial state of the Big Bang.

We can summarize some of our model's features (many of which are calculated later in this chapter) with:

1. The universe is a macroscopic object in terms of content and as such can be described by classical physics – modified by quantum effects.

2. Quantum fluctuations do not play any significant role in the Big Bang Epoch because of the nature of Two-Tier quantum field theory. For example, the quantum fluctuations of the quantized gravitational field were shown to be zero in chapter 7 of Blaha (2004):

$$<0|h_{\mu\nu}(X)h_{\alpha\beta}(X)|0> = \int d^3p \; b'_{\mu\nu\alpha\beta}(p) \; e^{-p^i p^j \Delta T_{ij}(0)}/[(2\pi)^3 2\omega_p] = 0 \quad (7.3.8.3.3)$$

3. All Two-Tier quantum fields also have zero quantum fluctuations. This behavior holds whether they are quantized in flat space or in a curved space. Thus quantum fluctuations (or foam) are not an issue in Two-Tier theories.

4. The radiation and matter in the Big Bang Epoch produces, in effect, a classical Y-quanta blackbody spectrum that modifies the nature of space making it complex with comparable real and imaginary spatial parts. This effect appears in the spatial scale factor of the Extended Robertson-Walker metric described in previous chapters. A Reality group transformation transforms spatial coordinates to real physical values.

5. The usual scale factor a(t) is determined by the conventional Standard Cosmological Model classical Einstein equation since its source is the macroscopic energy density.

6. The radius of the universe (with both real and imaginary parts) at the beginning of the Big Bang Epoch is

$$r_{universe}(0) = a_{BBRW}(0)/k^{1/2} = \gamma(1-i)/(2k^{1/2}) = 4.3(1-i) \times 10^{-65} \text{ cm} \quad (15.1.1)$$

The physical radius is the absolute value of $r_{universe}(0)$. The confinement of particles to a radius of this size means that they cannot be considered to be localized but, rather, they are spread over the entire volume of the universe.

7. The small size of the universe implies Two-Tier potentials between particles, and particle propagators, are effectively zero for all particles. Consequently the universe is in a metastable state. Some idea of the relative potential energy of Two-Tier QFT particles vs. ordinary QFT particles can be gleaned by comparing a standard Newton-Coulomb type of potential (knowing that it would be modified in strong gravitational fields if they were present)

$$V_{std} = 1/r \quad (15.1.2)$$

with the Two-Tier potential at short distances (below the Planck scale):

$$V_{tt} = 2\sqrt{\pi} \, M_c^2 r \quad (15.1.3)$$

obtained from eq. 7.3.9.3 of Blaha (2004). We have not displayed the coupling constant in eqs. 15.1.2 and 15.1.3. At $r = r_{universe}(0) = 4.3 \times 10^{-65}$ cm

$$V_{std} = 4.57 \times 10^{59} \text{ eV} \quad (15.1.4)$$

$$V_{tt} = 2\sqrt{\pi} \, M_c^2 r = 0.00115 \text{ eV} \quad (15.1.5)$$

Thus the Two-Tier potential between particles is negligible at distances up to the radius of the Big Bang Epoch universe at t = 0 since the Heisenberg Uncertainty Principle, when applied using the "diameter" of the universe as the uncertainty in position, implies the uncertainty in a particle's energy, and thus the scale of particle energies, is (coincidentally) of the order of 10^{59} eV. (See the discussion in the following sections.)

8. In view of the above points it is reasonable, and self-consistent, to use the Extended Robertson-Walker metric that we have developed in the preceding chapters.

9. The t = 0 universe is very dense with an energy density of the order of 10^{339} g/cm^3 of particles with negligibly small, non-singular (as particle separation goes to zero) interactions. (See the discussion below.)

10. The Big Bang Epoch is a metastable state. Due to the form of the Two-Tier gravitational potential it has a much higher gravitational potential energy at t = 0 than when the radius of the universe is about 10^{-33} cm. **In fact, Gravity is a repulsive force (anti-gravity!) at distances less than 9.08×10^{-34} cm.** (Fig. 7.3.9.3 of Blaha (2004) is reproduced on the next page for the reader's convenience.)

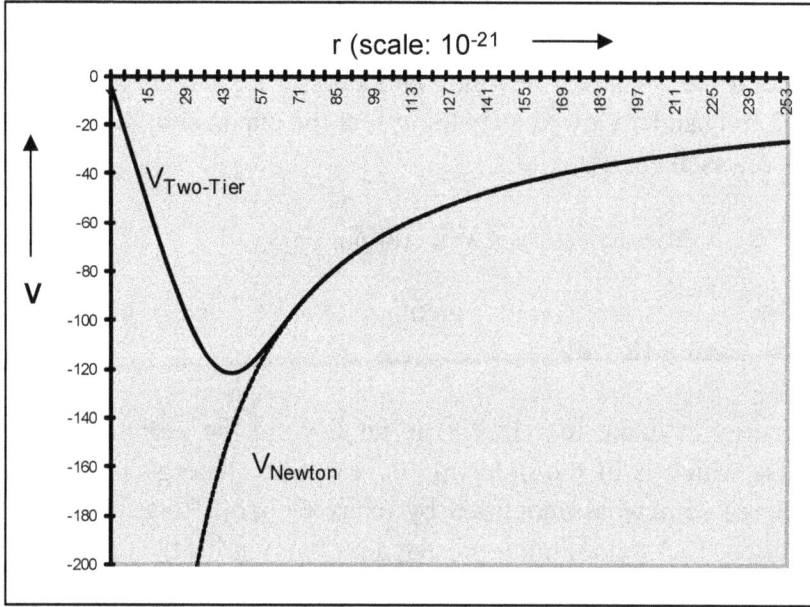

Figure 15.1.1. A plot of the Two-Tier gravitational potential (solid line) and the Newtonian gravitational potential (dashed line). Note anti-gravity at short distances. See Fig. 7.3.9.3 of Blaha (2004) for more details.

The Two-Tier gravitational potential between two particles has a minimum at (eq. 7.3.9.5 of Blaha (2004))

$$r_{MIN} = \pi^{-\frac{1}{2}}M_c^{-1} = 9.08 \times 10^{-34} \text{ cm} \tag{15.1.6}$$

At the minimum the gravitational $V_{Two-Tier}$ has the value:

$$V_{Two-TierMIN} = -.8427\sqrt{\pi} \, GM_c = -1.22 \times 10^{-28} \text{ eV}^{-1} \tag{15.1.7}$$

Since the energy of the universe is confined within a radius of 10^{-65} cm at $t = 0$ there is a tremendous release of gravitational potential energy as the universe expands to 10^{-33} cm and beyond. This energy is converted initially into kinetic energy and can be viewed as helping fuel the expansion. A crude approximation to this released energy is

$$\Delta E \sim (\rho V_{universe})^2\, V_{Two\text{-}TierMIN}$$

$$\cong (1.6 \times 10^{212}\ \text{g/cm}^3 \times 1.8 \times 10^{-100}\ \text{cm}^3)^2 \times 1.22 \times 10^{-28}\ \text{eV}^{-1}$$
$$\cong 3 \times 10^{262}\ \text{eV} \tag{15.1.8}$$

The energy estimate (eq. 15.1.8) is far beyond the current total energy of the universe which is of the order of 10^{57} eV. Since energy is not conserved (It is considered somewhat undefined by many General Relativists.) as the universe evolves eq. 15.1.8 does not represent a problem in itself.

11. The temperature of the real part of the universe at $t = 0$ is

$$\kappa T \approx 2 \times 10^{89}\ \text{eV} \tag{15.1.9}$$

as shown later.

15.2 The Radius of the Universe in the Big Bang Epoch

In this section we will consider the radius of the universe at the point of the Big Bang ($t = 0$) and its relatively slow growth during the Big Bang Epoch ($t \le t_c \cong 10^{-165}$ s).

The current radius[80] of the universe is 4.314×10^{28} cm. Motivated by that finding we assume the current radius of the universe in the Robertson-Walker model is:

Assumption: $r_{universe}(t_{now}) = a(t_{now})/k^{\frac{1}{2}} \approx 4.314 \times 10^{28}\ \text{cm}$ (15.2.1)

---

[80] M. Tanabashi *et al* (Particle Data Group) *op. cit.*

It is reasonable to define the radius of the universe at earlier times correspondingly. In our Two-Tier blackbody model for the universe it is

$$r_{universe}(t) = a_{BBRW}(t, \check{r} = 1)/k^{\frac{1}{2}} \qquad (15.2.2)$$

with

$$a_{BBRW}(t, \check{r}) = \{-i\gamma\check{r} + a(t) + [\gamma^2 - 2i\gamma\check{r}a(t) + a^2(t)]^{\frac{1}{2}}\}/2 \qquad (15.3.2.19)$$

Thus the radius of the universe is complex and given by

$$r_{universeBBRW}(t, \check{r}=1) = \{-i\gamma + a(t) + [\gamma^2 - 2i\gamma a(t) + a^2(t)]^{\frac{1}{2}}\}/(2k^{\frac{1}{2}}) \qquad (15.2.3)$$

The physical radius of the universe is the absolute value of eq. 15.2.3. It is calculated by applying a Reality group transformation.

The real and imaginary parts of $r_{universeBBRW}(t)$ are:

$$\text{Re } r_{universeBBRW}(t) = \{a(t) + [R(1 + \cos\psi)/2]^{\frac{1}{2}}\}/(2k^{\frac{1}{2}}) \qquad (15.2.4a)$$

and

$$\text{Im } r_{universeBBRW}(t) = \{-\gamma - [R(1 - \cos\psi)/2]^{\frac{1}{2}}\}/(2k^{\frac{1}{2}}) \qquad (15.2.4b)$$

where

$$R = [(\gamma^2 + a^2(t))^2 + 4\gamma^2 a^2(t)]^{\frac{1}{2}} \qquad (15.3.2.12)$$

and

$$\cos\psi = (\gamma^2 + a^2(t))/R \qquad (15.3.2.13)$$

From the behavior of $a_{BBRW}(t)$ displayed in Figs. 15.3.5.1 – 15.3.5.3 we see that it is well approximated by eqns. 15.3.2.24 – 15.3.2.29 with $\check{r} = 1$. Therefore for $t < t_c$

$t < t_c$

$$\text{Re } r_{universeBBRW}(0 \le t < t_c) \cong (\gamma + a(t))/(2k^{\frac{1}{2}}) \qquad (15.2.5a)$$

$$\text{Re } r_{universeBBRW}(0) \cong \gamma/(2k^{\frac{1}{2}}) = 4.278 \times 10^{-65} \text{ cm} \qquad (15.2.5b)$$

$$\text{Im } r_{universeBBRW}(0 \leq t < t_c) \cong -(\gamma + a(t))/(2k^{1/2}) \qquad (15.2.6a)$$

$$\text{Im } r_{universeBBRW}(0) \cong -\gamma/(2k^{1/2}) = -4.278 \times 10^{-65} \text{ cm} \qquad (15.2.6b)$$

from eqns. 15.3.2.24 and 15.3.2.25. Since the Planck length is 1.61×10^{-33} cm we see the real part of the radius of the universe at $t = 0$ (until $t \approx t_c$) is over thirty orders of magnitude smaller than the Planck length.

If we use eqns. 15.3.2.28 and 15.3.2.29 then the radius at present is

$t = t_{now}$

$$r_{universeBBRW}(t_{now}) = a(t_{now})/k^{1/2} - i\gamma/k^{1/2} \qquad (15.2.7)$$

with

$$\text{Re } r_{universeBBRW}(t_{now}) = a(t_{now})/k^{1/2} = 1.34 \times 10^{28} \text{ cm} \qquad (15.2.8)$$

as above (eq. 15.2.1), and

$$\text{Im } r_{universeBBRW}(t_{now}) = -\gamma/k^{1/2} = -8.55 \times 10^{-65} \text{ cm} \qquad (15.2.9)$$

The ratio of the current, and the $t = 0$ real parts of the, radius of the universe is huge:

$$r_{universe}(t_{now})/(\text{Re } r_{universeBBRW}(0)) = 1.008 \times 10^{93} \qquad (15.2.10)$$

The real part of the universe has expanded dramatically while the imaginary part has remained almost constant in size. The physical radius of the universe for $t < t_c$ and $t = t_{now}$ are the absolute values of the complex radius values above.

The value of the Robertson-Walker radius coordinate at points *within* the universe is specified by

$$r_{RW}(t, \check{r}) = a_{BBRW}(t, \check{r})r \qquad (15.2.11)$$

The coordinate r is related to the $\check{r}$ coordinate by

$$r = 2k^{-1/2}\check{r}(1 + \check{r}^2)^{-1} \qquad (10.4.11)$$

Thus

$$r_{RW}(t, \check{r}) = 2k^{-\frac{1}{2}} \check{r} a_{BBRW}(t, \check{r})(1 + \check{r}^2)^{-1} \qquad (15.2.12)$$

The real and imaginary parts of $a_{BBRW}(t, \check{r})$ are specified in eqns. 15.3.2.24 – 15.3.2.29. Again we note the absolute values of eqns. 15.2.11-15.2.12 are the physical values of the Extended Robertson-Walker radius coordinate.

15.2.1 Localization of Particles at t = 0

The radius of the universe in the neighborhood of t = 0 is approximately 4.278×10^{-65} cm. The particles within that incredibly small universe are still the "wave-particles" that we are familiar with within the framework of quantum mechanics. As such, the position and momentum of the particles must satisfy the Heisenberg Uncertainty Condition:

$$\Delta p \Delta x \geq \hbar \qquad (15.2.1.1)$$

where $\hbar$ is Planck's constant divided by 2π. In view of the extraordinary small size of the universe – much smaller than the Compton wavelength of any known massive particle – the "spread" (uncertainty) in a particle's position is set by the radius of the universe:

$$\Delta x \approx 2 \, |r_{universeBBRW}| \qquad (15.2.1.2)$$

Thus the "spread" in the momentum of a particle (the "size" of the region in momentum space where the Fourier transform of the particle's wave function is large) is

$$\Delta p \approx \hbar/\Delta x \approx \hbar/[2|\text{Re } r_{universeBBRW}|] = 1.6 \times 10^{59} \text{ eV} = 1.3 \times 10^{31} M_{Planck} \qquad (15.2.1.3)$$

One can only view the particles in the universe in the neighborhood of t = 0 as spread across the entire universe. They are entirely <u>un</u>localized within the universe from a quantum viewpoint. Since all forces are non-singular in the very small universe at the beginning of time we can view the particles as co-resident in the same spatial region that constitutes the universe. (Simply put, they interpenetrate each other.) Thus the question of particle horizons and the homogeneity of the universe in the Beginning are irrelevant. *The universe today does not scale down to a micro-universe of the same sort as our universe in the neighborhood of t = 0.*

15.3 Volume of Big Bang Metastate

Using the Extended Robertson-Walker model we find the total volume

$$V(t) = 2\pi^2 R^3(t) k^{-\frac{3}{2}}$$
$$= 2\pi^2 \, r_{universeBBRW}{}^3(t, \, \check{r}=1) \tag{15.3.1}$$

Earlier we found

$$r_{universeBBRW}(t, \, \check{r}=1) = \{-i\gamma + a(t) + [\gamma^2 - 2i\gamma a(t) + a^2(t)]^{\frac{1}{2}}\}/(2k^{\frac{1}{2}}) \tag{15.2.3}$$

Therefore

$$V(0) = \pi^2 \gamma^3 |(1-i)|^3/(4k^{3/2}) \tag{15.3.2}$$
$$= 4.37 \times 10^{-192} \text{ cm}^3$$

and

$$V(t_c) = \pi^2 \gamma^3 |1 - i + \sqrt{2}(1-i)^{\frac{1}{2}}|^3/(4k^{3/2}) \tag{15.3.3}$$
$$= \pi^2 \gamma^3 |1 - i + \sqrt{2}(1-i)^{\frac{1}{2}}|^3/(4k^{3/2})$$
$$= 2.6398 \times 10^{-191} \text{ cm}^3$$

using eq. 13.3.2.14.

15.4 Big Bang Epoch Total Y-Field Expansion Energy of Universe

We can estimate the total Y-field expansion energy[81] of the universe at t= 0 and t = t_c using

---

[81] In the Big Bang metastate the masses of particles are zero. The interparticle interactions are also zero. Radiation pressure is zero as well. See section 15.1 and Fig. 15.1.1 in particular. Thus only Y-field black body radiation causes expansion. Its energy is calculated in this section.

$$\rho_{BBRW} (t = 0, \v{r} = 0) \cong (3/8\pi G)[4H_0\Omega_{,\gamma}^{1/2}/\gamma^2 + k] \cong (3/2\pi\gamma^2 G)H_0\Omega_{,\gamma}^{1/2} \qquad (13.7.3)$$
$$\cong 2.9 \times 10^{194} \text{ gm/cm}^3 = 1.63 \times 10^{218} \text{ GeV/cm}^3$$

$$\rho_{BBRW} (t = 0, \v{r}=1) \cong (3/8\pi G)[2H_0\Omega_{,\gamma}^{1/2}/\gamma^2 + k] \cong (3/4\pi\gamma^2 G)H_0\Omega_{,\gamma}^{1/2} \qquad (13.7.4)$$
$$\cong 1.45 \times 10^{194} \text{ gm/cm}^3 = 8.16 \times 10^{217} \text{ GeV/cm}^3$$

$$\rho_{BBRW} (t = t_c, \v{r} = 0) \cong 3/(8\pi G)[H_0\Omega_{,\gamma}^{1/2}/\gamma^2 + k] \cong 3/(8\pi\gamma^2 G)H_0\Omega_{,\gamma}^{1/2} \qquad (13.7.5)$$
$$\cong 7.26 \times 10^{193} \text{ gm/cm}^3 = 4.08 \times 10^{217} \text{ GeV/cm}^3$$

$$\rho_{BBRW} (t = t_c, \v{r}=1) \cong 3/(8\pi G)[H_0\Omega_{,\gamma}^{1/2}/(2\gamma^2) + k] \cong 3/(16\pi\gamma^2 G)H_0\Omega_{,\gamma}^{1/2} \qquad (13.7.6)$$
$$\cong 3.63 \times 10^{193} \text{ gm/cm}^3 = 2.04 \times 10^{217} \text{ GeV/cm}^3$$

At $t = 0$ we estimate the total energy using the average:

$$\rho_{BBRWaverage} (t = 0) = (\rho_{BBRW} (t = 0, \v{r} = 0) + \rho_{BBRW} (t = 0, \v{r}=1))/2$$
$$= 12.23 \times 10^{217} \text{ GeV/cm}^3 \qquad (15.3.4)$$

and the volume from eq. 15.3.2

$$V(0) = 4.37 \times 10^{-192} \text{ cm}^3 \qquad (15.3.5)$$

to obtain the total Y-field expansion energy

$$E_{BB}(0) = 5.34 \times 10^{35} \text{ eV} \qquad (15.3.6)$$

At $t = t_c$ we estimate the total energy using the average:

$$\rho_{BBRWaverage} (t = t_c) = (\rho_{BBRW} (t = t_c, \v{r} = 0) + \rho_{BBRW} (t = t_c, \v{r}=1))/2$$
$$= 3.06 \times 10^{217} \text{ GeV/cm}^3 \qquad (15.3.7)$$

and the volume from eq. 15.3.3

$$V(t_c) = 2.64 \times 10^{-191} \text{ cm}^3 \qquad (15.3.8)$$

to obtain the total Y-field expansion energy

$$E_{BB}(t_c) = 8.07 \times 10^{35} \text{ eV} \qquad (15.3.9)$$

The difference between eqs. 15.3.6 and 15.3.9 can be attributed to round off error associated with averaging the densities.

15.5 A Quantum Big Bang and Evolutionary Theory

At this point we have shown that our Quantum Big Bang theory does not have a singularity at the beginning of the universe. It exhibits the known behavior of the universe since 380,000 years after the Big Bang epoch. Thus the long sought inflaton field turns out to be the quantum field, $Y^\mu(y)$, appearing in our definition of quantum coordinates. Remarkably our quantum coordinates also eliminate the divergences that have plagued quantum field theory for almost eighty years and enable calculations in The Extended Standard Model and Quantum Gravity to be divergence free without the cumbersome renormalization techniques that were needed for ElectroWeak theory and would not work for Quantum Gravity.

So our quantum coordinates free The Extended Standard Model and Quantum Gravity of infinities, both now, and at the beginning of the universe. Naturally this happy removal of infinities through one simple mechanism is a remarkable result that, we believe, reflects the simplicity of Nature when properly understood.Leibniz's Minimax Principle is realized by Two-Tier quantum theory, which removes the infinities in The Extended Standard Model, Quantum Gravity, and at the beginning of the universe.

16. Summary of the Influx-Less Epochs Case

The Expansion of the universe has a Radiation-dominated epoch and a Matter-dominated epoch. In this chapter we will summarize relevant quantities in these epochs as well as the Explosive and Big Bang epochs *assuming* that there is no influx of mass-energy from the Megaverse. Fig. 16.1 displays the data in a table format.

16.1 The Four Epochs

t	Time	Re a(t)	Im a(t)	a(t, ř)
t_{now}	13.8 Gyr	1	-6.38×10^{-93}	
Explosive			↑	$\exp(H_0\Omega_\Lambda^{\frac{1}{2}} t)$
t_E	9.1 Gyr	,76		
Matter				$2H_0\Omega_m^{\frac{1}{2}} t)^{2/3}$
t_{RM}	2.2×10^{-5} Gyr	1.8×10^{-93}		
Radiation				$(2H_0\Omega_\gamma^{\frac{1}{2}} t)^{\frac{1}{2}}$
t_c	1.26×10^{-165} s	6.38×10^{-93}		
Big Bang metastate				Re $a_{BBRW}(t, ř) = (\gamma + a(t))/2$ Im $a_{BBRW}(t, ř) = (\gamma + a(t)) ř/2$
0	0 s	3.19×10^{-93}	-3.19×10^{-93}	
?				

Figure 16.1 Times and scale factors of the four epochs.

16.2 Big Bang Metastate Data Summary

Fig. 16.2 summarizes the results for the Big Bang metastate and the radiation-dominated–matter-dominated boundary t_{RM}.

Time	0	t_c
Phase	Big Bang Beginning	Metastable End
Time	0	1.26×10^{-165}
Re a(t)	3.19×10^{-93}	8.166×10^{-93}
Im a(t)	-3.16×10^{-93}	-5.24×10^{-93}
Re radius	4.278×10^{-65} cm	8.5×10^{-65} cm
Im radius	4.278×10^{-65} cm	8.5×10^{-65} cm
Volume	4.37×10^{-192} cm^3	2.6398×10^{-191} cm^3
Central Expansion Energy Density	1.63×10^{218} GeV/cm^3	4.08×10^{217} GeV/cm^3
Edge Expansion Energy Density	8.16×10^{217} GeV/cm^3	2.04×10^{217} GeV/cm^3
Total Expansion Energy[82]	5.34×10^{35} eV	8.07×10^{35} eV
Hubble Constant[83] (km s^{-1} Mpc^{-1})	1.79×10^{218}	1.14×10^{126}

Figure 16.2 The Big Bang Epoch detailed data.

---

[82] The expansion energy does not include the mass-energy of particles in the Big Bang universe. It only includes y field black body energy which drives the initial expansion. All particles are massless initially and all interparticle forces are zero.

[83] In the Big Bang metastable state we display the Hubble constant at the "expanding' edge.

16.3 Some Background Equations for Big Bang Metastate Data

Below are some of the equations supporting the data in Fig. 16.2.

$\check{r} \in [0, 1]$ From center to edge of Big Bang Metastable universe

$0 \leq t \leq t_c$

$$\text{Re } a_{BBRW}(t, \check{r}) \cong \gamma/2 + a(t)/2 \qquad (13.3.2.24)$$

$$\text{Im } a_{BBRW}(t, \check{r}) \cong -\gamma\check{r}/2 - a(t)\check{r}/2 \qquad (13.3.2.25)$$

$t_c < t$

$$\text{Re } a_{BBRW}(t, \check{r}) \cong a(t)\{1 + (1 + \check{r}^2)\gamma^2/4\} \cong a(t) \qquad (13.3.2.26)$$

$$\text{Im } a_{BBRW}(t, \check{r}) \cong -\gamma\check{r}/2 - [\check{r}^2 + \gamma^2/(4a^2(t))]^{\frac{1}{2}}\gamma/2 \qquad (13.3.2.27)$$

$t \to \infty$

$$\text{Re } a_{BBRW}(t, \check{r}) \cong a(t)\{1 + (1 + \check{r}^2)\gamma^2/4\} \cong a(t) \qquad (13.3.2.28)$$

$$\text{Im } a_{BBRW}(t, \check{r}) \cong -\gamma\check{r} = -8.166 \times 10^{-93}\check{r} \qquad (13.3.2.29)$$

$0 \leq t < t_c$

$$\kappa T_< \equiv \kappa T(0 \leq t < t_c) \cong 2\kappa T_0(1 + i\check{r})/\chi = 4\pi^{-3/2} M_c(1 + i\check{r})$$
$$= 2.72 \times 10^{29}(1 + i\check{r}) \text{ eV} \qquad (13.3.3.4)$$
$$= 3.155 \times 10^{33}(1 + i\check{r}) \,^\circ \text{K}$$

$t_c < t$

$$\kappa T_> \equiv \kappa T(t_c < t)$$
$$\cong \kappa T_0(a(t) + i\gamma\check{r})/a^2(t) = \kappa T_0/a(t) \qquad (13.3.3.5)$$

$0 \leq t < t_c$

$$H(t, \check{r}) = H_{BBRW}(t, \check{r}) \cong 4H_0\Omega_\gamma^{\frac{1}{2}}|[1 - \check{r}^2 + 2i\check{r}]/[(\gamma + a(t))^2(1 + \check{r}^2)^2]| \qquad (13.4.5)$$

$$H_{BBRW}(0, \check{r}) \cong 4H_0\Omega_\gamma^{\frac{1}{2}}|[1 - \check{r}^2 + 2i\check{r}]/[\gamma^2(1 + \check{r}^2)^2]| \qquad (13.4.6)$$

$$H_{BBRW}(0, 1) \cong 2H_0\Omega_\gamma^{\frac{1}{2}}/\gamma^2 = 5.8 \times 10^{198} \text{ s}^{-1} \qquad (13.4.7)$$
$$= 1.79*10^{218} \text{ km s}^{-1} \text{ Mpc}^{-1}$$

$$H_{BBRW}(0, 0) \cong 4H_0\Omega_\gamma^{\frac{1}{2}}1/\gamma^2 = 2H_{BBRW}(0, 1) \qquad (13.4.8)$$

$$H_{BBRW}(t_c, 1) \cong 4H_0\Omega_\gamma^{1/2}|i|/[2(\gamma + a(t_c))^2|] \qquad (13.4.9)$$
$$= 2\ H_0\Omega_\gamma^{1/2}/(\gamma + \gamma^2) \cong 2\ H_0\Omega_\gamma^{1/2}/\gamma = 3.7 \times 10^{106}\ s^{-1}$$
$$= 1.14*10^{126}\ km\ s^{-1}\ Mpc^{-1}$$

$$H_{BBRW}(t_c, 0) \cong H_0\Omega_\gamma^{1/2}/\gamma^2 = 2.9 \times 10^{198}\ s^{-1} \qquad (13.4.10)$$
$$= 8.95*10^{217}\ km\ s^{-1}\ Mpc^{-1}$$

$$r_{universeBBRW}(t) = \{-i\gamma + a(t) + [\gamma^2 - 2i\gamma a(t) + a^2(t)]^{1/2}\}/(2k^{1/2}) \qquad (15.2.3)$$

$$Re\ r_{universeBBRW}(t) = \{a(t) + [R(1 + \cos\psi)/2]^{1/2}\}/(2k^{1/2}) \quad (15.2.4a)$$

$$Im\ r_{universeBBRW}(t) = \{-\gamma - [R(1 - \cos\psi)/2]^{1/2}\}/(2k^{1/2}) \quad (15.2.4b)$$

$$R = [(\gamma^2 + a^2(t))^2 + 4\gamma^2 a^2(t)]^{1/2} \qquad (15.3.2.12)$$
$$\cos\psi = (\gamma^2 + a^2(t))/R \qquad (15.3.2.13)$$

$$Re\ r_{universeBBRW}(0 \le t < t_c) \cong (\gamma + a(t))/(2k^{1/2}) \qquad (15.2.5a)$$
$$Re\ r_{universeBBRW}(0) \cong \gamma/(2k^{1/2}) = 4.278 \times 10^{-65}\ cm \qquad (15.2.5b)$$
$$Im\ r_{universeBBRW}(0 \le t < t_c) \cong -(\gamma + a(t))/(2k^{1/2}) \qquad (15.2.6a)$$
$$Im\ r_{universeBBRW}(0) \cong -\gamma/(2k^{1/2}) = -4.278 \times 10^{-65}\ cm \qquad (15.2.6b)$$

$$r_{universeBBRW}(t_{now}) = a(t_{now})/k^{1/2} - i\gamma/k^{1/2} \qquad (15.2.7)$$
$$Re\ r_{universeBBRW}(t_{now}) = a(t_{now})/k^{1/2} = 1.34 \times 10^{28}\ cm \qquad (15.2.8)$$
$$Im\ r_{universeBBRW}(t_{now}) = -\gamma/k^{1/2} = -8.55 \times 10^{-65}\ cm \qquad (15.2.9)$$

$$\rho_{BBRW}\ (t = 0, \check{r} = 0) \cong (3/8\pi G)[4H_0\Omega_\gamma^{1/2}/\gamma^2 + k] \cong (3/2\pi\gamma^2 G)H_0\Omega_\gamma^{1/2} \qquad (13.7.3)$$
$$\cong 2.9 \times 10^{194}\ gm/cm^3 = 1.63 \times 10^{218}\ GeV/cm^3$$

$$\rho_{BBRW}\ (t = 0, \check{r}=1) \cong (3/8\pi G)[2H_0\Omega_\gamma^{1/2}/\gamma^2 + k] \cong (3/4\pi\gamma^2 G)H_0\Omega_\gamma^{1/2} \qquad (13.7.4)$$
$$\cong 1.45 \times 10^{194}\ gm/cm^3 = 8.16 \times 10^{217}\ GeV/cm^3$$

$$\rho_{\text{BBRW}} \,(t = t_c, \, \check{r} = 0) \cong 3/(8\pi G)[H_0\Omega_\gamma^{1/2}/\gamma^2 + k] \cong 3/(8\pi\gamma^2 G)H_0\Omega_\gamma^{1/2} \qquad (13.7.5)$$
$$\cong 7.26 \times 10^{193} \,\text{ gm/cm}^3 = \, 4.08 \times 10^{217} \,\text{ GeV/cm}^3$$

$$\rho_{\text{BBRW}} \,(t = t_c, \, \check{r}=1) \cong 3/(8\pi G)[H_0\Omega_\gamma^{1/2}/(2\gamma^2) + k] \cong 3/(16\pi\gamma^2 G)H_0\Omega_\gamma^{1/2} \qquad (13.7.6)$$
$$\cong 3.63 \times 10^{193} \,\text{ gm/cm}^3 = \, 2.04 \times 10^{217} \,\text{ GeV/cm}^3$$

17. Conventional Universe Expansion with an Influx from the Megaverse

This chapter calculates the mass-energy influx from the Megaverse and the consequent change in energy density with time. As a result of the influx the character of the epochs changes and the Hubble Constant becomes time dependent as current experiment seems to indicate.

17.1 Initial State for Influx Initiation

If we examine the time periods of the universe epochs in the absence of a Megaverse influx, then it appears that the beginning of the influx can be set to a time within the Matter-dominated epoch.[84]

Because available data[85] gives Hubble Constant values for the early universe at time t = t_T = 380,000 years = 1.19837 × 10^{13} s = 1.205 × 10^{-11} Gyr, a time within the Matter-dominated epoch we will calculate the change in the universe since that time due to the influx between then and now. Note that the 380,000 years before t_T is negligible compared to the 13.8 Gyr existence of the universe. Consequently the influx before t_T will be negligible compared to the influx in the 13.4 Gyr period afterwards.

We find the scale factor at t_T = 380,000 years to be

$$a(t_T) = (2H_0\Omega_\gamma^{1/2} t_T)^{1/2} \qquad (17.1)$$
$$= 0.0066 \qquad (17.2)$$

---

[84] See Fig. 16.1.
[85] See, for example, CMB data of K. Aylor *et al*, arXiv:1811.00537v1 (2018) and similar studies.

17.2 Megaverse Influx Equation for the Energy Density

In chapter 14 we determined the time-dependent equation for the Megaverse influx:

$$d\Omega_T(t)/dt = C_M\, k^{1/2}[\Omega_H(t)+ \Omega_T(t)]\, /a(t) \tag{14.4}$$

The solution to eq. 14.4 is[86]

$$\Omega_T(t) = C_M k^{1/2} e^{F(t)} \int_{t_T}^{t} dt''\, e^{-F(t'')}\Omega_H(t'')/a(t'') \tag{14.5}$$

where

$$F(t) = C_M k^{-1/2} \int_{t_T}^{t} dt'/a(t') \tag{14.2}$$

17.3 Approximate Solution Based on Matter-Dominated a(t)

We can approximate eq.14.4 based on the assumption that $\Omega_H(t) \gg \Omega_T(t)$. This assumption seems reasonable since the increasing expansion rate of the universe is relatively modest. We attribute the "modestly" increasing rate to a modest mass-energy influx from the Megaverse. Its relative "smallness" is due to the dominance of $\Omega_H(t)$ over the incremental mass-energy factor $\Omega_T(t)$. Therefore we approximate

$$d\Omega_T(t)/dt = C_M\, k^{1/2}\Omega_H(t)/a(t) \tag{17.3}$$

Eq. 17.3 can be solved:[87]

$$\Omega_T(t) = C_M k^{1/2} \int_{t_T}^{t} dt'\Omega_H(t')/a(t') \tag{17.4}$$

---

[86] This approximation causes the energy influx to not "feed" on itself after it enters our universe. Thus, to some extent, it limits the growth of the energy density and the Hubble Constant..

[87] This approximation yields a lower value for the total energy density. As a result da/dt will be smaller and the Hubble constant obtained will be lower.

Combining eq. 17.4 with the Einstein equation:

$$\dot{a}^2(t) - 8\pi G\rho_{tot}a^2(t)/3 = -k \tag{11.19}$$

we have an implicit equation for a(t) in the Megaverse mass-energy influx epoch, which continues up to the present day:

$$\dot{a}^2(t) - (8\pi G/3)\, a^2(t)\rho_{crit}(\Omega_H(t) + \Omega_T(t)) = -k \tag{17.5}$$

where

$$\Omega_H(t) = \Omega_\gamma(t) + \Omega_m(t) + \Omega_d(t) + \Omega_\Lambda(t) \tag{9.10}$$

with

$$\Omega_\gamma(t) = 5.38 \times 10^{-5}/a^4(t) \tag{17.6}$$
$$\Omega_m(t) = 0.046/a^3(t)$$
$$\Omega_d(t) = 0.268/a^3(t)$$
$$\Omega_\Lambda(t) = 0.692$$
$$\Omega_M(t) = \Omega_m(t) + \Omega_d(t) = 0.308/a^3(t) \tag{17.7}$$

The dominant terms in $\Omega_H(t)$ are the sum of the baryon density, the cold dark matter density, and the dark energy density:

$$\Omega_H(t) \approx \Omega_M(t) + \Omega_\Lambda = \Omega_M/a^3(t) + \Omega_\Lambda \tag{17.8}$$

where

$$\Omega_M = 0.308 \tag{17.8a}$$

Thus

$$\Omega_T(t) = C_M k^{\frac{1}{2}} \int_{t_T}^{t} dt'[\Omega_M/a^4(t') + \Omega_\Lambda/a(t')] \tag{17.9}$$

Combining the above equations we obtain

$$da/dt \approx H_0 [\Omega_M /a(t) + \Omega_\Lambda a^2(t) + C_M k^{\frac{1}{2}} a^2(t) \int_{t_T}^{t} dt' (\Omega_M/a^4(t') + \Omega_\Lambda/a(t'))]^{\frac{1}{2}} \qquad (17.10)$$

To calculate $a(t)$ we will assume that it has the same form as the Matter dominated epoch in chapter 13 since we use $t_T = 380,000$ years[88] to allow direct comparison with astrophysical Hubble Constant data:
:

$$a(t) = \beta t^\gamma \qquad (17.11)$$

with the normalization

$$a(t_{now}) = 1 \qquad (17.12)$$

resulting in

$$\beta = t_{now}^{-\gamma} \qquad (17.12a)$$

Then after a time integration eq. 17.10 becomes

$$a(t) - a(t_T) \approx H_0 \int_{t_T}^{t} dt' [\Omega_M/(\beta t'^\gamma) + \Omega_\Lambda(\beta t'^\gamma)^2 + d_M[\Omega_M \beta^{-2}(t'^{1-2\gamma} - t'^{2\gamma} t_T^{1-4\gamma})/(1-4\gamma) +$$
$$+ \Omega_\Lambda \beta(t'^{1+\gamma} - t_T t'^{1-\gamma} t'^{2\gamma}/(1-\gamma))]^{\frac{1}{2}} \qquad (17.13)$$

where

$$d_M = C_M k^{\frac{1}{2}} \qquad (17.14)$$

Note eq. 17.11 and 17.13 imply

$$\Omega_T(t) = d_M [\Omega_M \beta^{-4}(t^{1-4\gamma} - t_T^{1-4\gamma})/(1-4\gamma) + \Omega_\Lambda(t^{1-\gamma} - t_T^{1-\gamma})/(\beta(1-\gamma))] \qquad (17.9a)$$

Since the second and following terms on the right (the influx increment) in eq. 17.13 are presumably[89] small compared to the first term we approximate eq. 17.13 as

---

[88] This time is within the matter-dominated epoch using the conventional $a(t)$ of chapter 13. Note that the 380,000 years before $t = t_{cSE}$ is negligible compared to the 13.8 Gyr existence of the universe. Cosequently the influx before t_{cSE} will be negligible compared to the influx in the 13.4 Gyr period afterwards.
[89] We check this in section 17.5.

$$a(t)-a(t_T) \approx H_0 \int_{t_T}^{t} dt' \sqrt{\Omega_M}\, (\beta t'^\gamma)^{-\frac{1}{2}} [1 + \tfrac{1}{2}\, (\Omega_\Lambda/\Omega_M)(\beta t'^\gamma)^3 + \tfrac{1}{2}(\beta t'^\gamma/\Omega_M)\, \Omega_T] \tag{17.15}$$

$$= H_0 \int dt' [\sqrt{\Omega_M}\, (\beta t'^\gamma)^{-\frac{1}{2}} + \tfrac{1}{2}\, \sqrt{\Omega_M}\, (\beta t'^\gamma)^{-\frac{1}{2}}\, (\Omega_\Lambda/\Omega_M)(\beta t'^\gamma)^3 + \tfrac{1}{2}\Omega_M^{-\frac{1}{2}}\beta^{\frac{1}{2}} t'^{\gamma/2}\, d_M \cdot$$
$$\cdot [\Omega_M\beta^{-4}(t^{1-4\gamma} - t_T^{1-4\gamma})/(1-4\gamma) + \Omega_\Lambda(t^{1-\gamma} - t_T^{1-\gamma})/(\beta(1-\gamma))]\,]$$

$$= H_0\, \beta^{-\frac{1}{2}}\Omega_M^{1/2}\, \{(t^{1-\gamma/2} - t_T^{1-\gamma/2})/(1-\gamma/2) +$$
$$+\, \tfrac{1}{2}\, \beta^3(\Omega_\Lambda/\Omega_M)(t^{2.5\gamma+1} - t_T^{2.5\gamma+1})/(2.5\gamma+1) +$$
$$+\, \tfrac{1}{2}\beta\Omega_M^{-1}d_M\{\Omega_M\beta^{-4}(t^{2-3.5\gamma}/(2-3.5\gamma) -$$
$$-\, t_T^{1-4\gamma}\, t^{\gamma/2+1}/(1+\gamma/2))/(1-4\gamma) +$$
$$+\, \Omega_\Lambda(t^{-\gamma/2+2}/(2-\gamma/2) - t_T^{1-\gamma}\, t^{\gamma/2+1}/(1+\gamma/2))/(\beta(1-\gamma)) -$$

$$-\, \tfrac{1}{2}\beta\Omega_M^{-1}d_M\{\Omega_M\beta^{-4}(t_T^{2-3.5\gamma}\, (1/(2-3.5\gamma) - 1/(1+\gamma/2))/(1-4\gamma) +$$
$$+\, \Omega_\Lambda(t_T^{-\gamma/2+2}(1/(2-\gamma/2) - 1/(1+\gamma/2))/(\beta(1-\gamma))\}$$

with $a(t_T) = 0.0066$ by eq. 17.2 where $t_T = 380{,}000$ years $= 1.19837 \times 10^{13}$ s.

Upon integration we obtain

$$a(t)-a(t_T) \approx H_0\beta^{-\frac{1}{2}}\sqrt{\Omega_M}\, (V(t) - V(t_T)) \tag{17.16}$$

where $a(t_T)$ is the matter-dominated scale factor value at time t_T and

$$V(t) = t^{1-\gamma/2}/(1-\gamma/2) + \tfrac{1}{2}\beta^3(\Omega_\Lambda/\Omega_M)t^{2.5\gamma+1}/(2.5\gamma+1) + \tag{17.17}$$
$$+\, \tfrac{1}{2}\, d_M\{\beta^{-3}\, (t^{2-3.5\gamma}/(2-3.5\gamma) - t_T^{1-4\gamma}\, t^{\gamma/2+1}/(1+\gamma/2))/(1-4\gamma) +$$
$$+\, (\Omega_\Lambda/\Omega_M)(t^{-\gamma/2+2}/(2-\gamma/2) - t_T^{1-\gamma}\, t^{1+\gamma/2}/(1+\gamma/2))/(1-\gamma)\}$$

The derivative of $a(t)$ is

$$da(t)/dt \approx H_0\beta^{-\frac{1}{2}}\sqrt{\Omega_M}\, dV(t)/dt \tag{17.16a}$$

Since we assume that the influx density is much smaller than the matter energy density in the matter-dominated phase during which the influx occurs we will use the matter epoch a(t) where $\gamma = 2/3$ and $\beta = t_{now}^{-\gamma}$, and treat the influx contribution as based on the total matter energy density. Then we see that V(t) becomes[90]

$$V_M(t) = 3t^{2/3}/2 + (3/16) t_{now}^{-2}(\Omega_\Lambda/\Omega_M)t^{8/3} + \tag{17.18}$$
$$+(9/2) d_M\{t_{now}^2(t^{-1/3} + t_T^{-5/3} t^{4/3}/4)/5 +$$
$$+ (\Omega_\Lambda/\Omega_M)(t^{5/3}/5 - t_T^{1/3} t^{4/3}/4)\}$$

and

$$V_M(t_T) = 3t^{2/3}/2 + (3/16) t_{now}^{-2}(\Omega_\Lambda/\Omega_M)t^{8/3} + \tag{17.18a}$$
$$+(9/8) d_M\{t_{now}^2 t_T^{-1/3} - (1/20) (\Omega_\Lambda/\Omega_M) t_T^{5/3}\}$$

Note the first term $t^{2/3}$ leading dependence of a(t) as expected from the choice of γ. We will see that the influx terms of eq. 17.18 causes a(t) and the Hubble parameter ("constant") to increase with time as seems to be the findings of astrophysical studies.

We define

$$a_M(t) = H_0\beta^{-\frac{1}{2}} \sqrt{\Omega_M} (V_M(t) - V_M(t_T)) + a(t_T) \tag{17.19}$$

where $a(t_T)$ is the matter-dominated scale factor value at time t_T and the derivative

$$da_M(t)/dt \approx H_0\beta^{-\frac{1}{2}} \sqrt{\Omega_M} dV_M(t)/dt \tag{17.19a}$$

where

$$dV_M(t)/dt = t^{-1/3} + \frac{1}{2} t_{now}^{-2}(\Omega_\Lambda/\Omega_M)t^{5/3} + \tag{17.20}$$
$$+(3/2)d_M\{t_{now}^2(-t^{-4/3} + t_T^{-5/3} t^{1/3})/5 + (\Omega_\Lambda/\Omega_M)(t^{2/3} - t_T^{1/3} t^{1/3})\}$$

---

[90] The coefficient of the $t^{2/3}$ term in a(t) expression inf eqs. 17.16 and 17.18 differs from the standard term:

$$a(t) \cong [2H_0\Omega_m^{\frac{1}{2}}t]^{2/3} \tag{13.2.3.8}$$

by a factor of 1.19 more due to the difference between the algebraic approximation giving eq. 13.2.3.8 and the above integration. This difference "washes out" when the Hubble Constant is calculated H = d(ln a)/dt.

We will use it in section 17.6 to determine the time dependence of the Hubble parameter (Constant) $H_M(t)$:

17.4 Evaluation of d_M and the Determination of C_T and ρ_{Mega}

The parameter d_M above can be expressed in terms of the universe's surface tension and G using eqs. 9.8 and 17.14 which give

$$d_M = C_T k^{1/2} G \rho_{Mega} \tag{17.21}$$

Since d_M must have the dimension s^{-1} by eq. 17.18, we define

$$C_T = C_{T1} \, ck^{-1} \tag{17.21a}$$

making C_{T1} a dimensionless constant where c is the speed of light. Then

$$\begin{aligned} d_M &= C_{T1} \, ck^{-1/2} G \rho_{Mega} \\ &= C_{T1} \, ck^{-1/2} G \rho_{crit} \rho_{M0} \end{aligned} \tag{17.22}$$

where C_{T1}, and ρ_{M0} are dimensionless constants, with

$$\rho_{Mega} = \rho_{M0} \, \rho_{crit} \tag{17.23}$$

We express ρ_{Mega} in terms of the multiple ρ_{M0} of our universe's critical density ρ_{crit}. C_{T1} is a surface tension constant which we will set to one in this chapter and the following chapters[91]

$$C_{T1} = 1 \tag{17.24}$$

---

[91] The value of C_{T1} probably should have a different constant value. When we later fit the model to the data to compare with experiment only the value of $C_{T1} \rho_{M0}$ will be fixed. We will suggest $C_{T1} = (L_{Planck} k^{1/2})^{-1} \approx 10^{62}$ where L_{Planck} is the Planck length since we will find $C_{T1} \rho_{M0}$ is of the order of 10^{-8}. Thus the Megaverse density factor ρ_{M0} will be of the order of 10^{-70}. (A physically reasonable value since a universe is a much higher density region of the Megaverse.) The choice of the dimensionless constant $C_{T1} = (L_{Planck} k^{1/2})^{-1}$ is reasonable since both factors in it refer to the curvature of the universe. Surface Tension depends on the radius of curvature.

Eq. 17.24 implies

$$d_M = 2.57 \times 10^{-19} \text{ s}^{-1} \rho_{M0} \qquad (17.25)$$

17.5 Megaverse Energy Density Bound

Since $\Omega_M = 0.308$ and $\Omega_\Lambda = 0.692$, we expect $\Omega_T(t_{now})$ to be of the order of 1 or less giving an *upper* bound on ρ_{M0} of 10^{-6}. The Megaverse density, $\rho_{Mega} = \rho_{M0} \rho_{crit}$, is thus very much less than our universe's critical energy density ρ_{crit}. If the Megaverse measure of density $\Omega_T(t_{now})$ is of the order of magnitude of Ω_M, then we expect $\Omega_T(t_{now})$ to be between 0 – 1, with the result the order of magnitude of ρ_{M0} is between 0 and 10^{-6} based on eq. 17.24. Since we wish to keep $\Omega_T(t_{now}) < \Omega_M$ for physical reasons, we will examine values of $C_{T1} \rho_{M0}$ below 10^{-6} in chapter 18.

17.6 The Time Dependent Hubble Constant H(t)

The time dependent Hubble "Constant" can be expressed in terms of $V_M(t)$ and its derivative as

$$H_M(t) = da_M(t)/dt / a_M(t) = dV_M(t)/dt/[\ V_M(t) - V_M(t_T) + (\ H_0 \beta^{-\frac{1}{2}} \sqrt{\Omega_M})^{-1} a(t_T)] \qquad (17.26)$$

using eqs. 17.18 – 17.20. In chapter 20 we will use an alternate, more convenient, form for H(t):

$$H(t) = (da/dt)/a(t) = [H_0^2 \rho_{tot}(t)/\rho_{cri} - c^2 k/a^2(t)]^{\frac{1}{2}} \qquad (20.1)$$

18. "Conventional" Influx Megaverse Mass-Energy Accretion in our Universe

In this chapter we determine the time dependence of the Megaverse mass-energy influx as a function of time and its impact on the density of the universe.

18.1 Influx Mass-Energy Density $\rho_T(t)$ as a Function of Time

Eq. 17.9a implies the influx mass-energy density from the Megaverse is

$$\rho_T(t) = \rho_{crit}\Omega_T(t) \qquad\qquad (18.1)$$

where

$$\Omega_T(t) = d_M\,[\Omega_M\beta^{-4}(t^{1\,-\,4\gamma} - t_T^{\,1\,-\,4\gamma})/(1 - 4\gamma) + \Omega_\Lambda(t^{1\,-\,\gamma} - t_T^{\,1\,-\,\gamma})/(\beta(1 - \gamma))] \qquad (17.9a)$$

Eq. 18.1 is plotted in Fig. 18.1. Note $\Omega_T(t)$ is approximately constant. The values of ρ_{M0} and the average value of $\Omega_T(t)$ are

ρ_{M0}	Average $\Omega_T(t)$
3.7037E-08	0.0304866
7.40741E-08	0.0609732
1.48148E-07	0.1219463
2.96296E-07	0.2438927
5.92593E-07	0.4877854
1.18519E-06	0.9755708

Table 18.1 Megaverse Density Factors and average $\Omega_T(t)$.

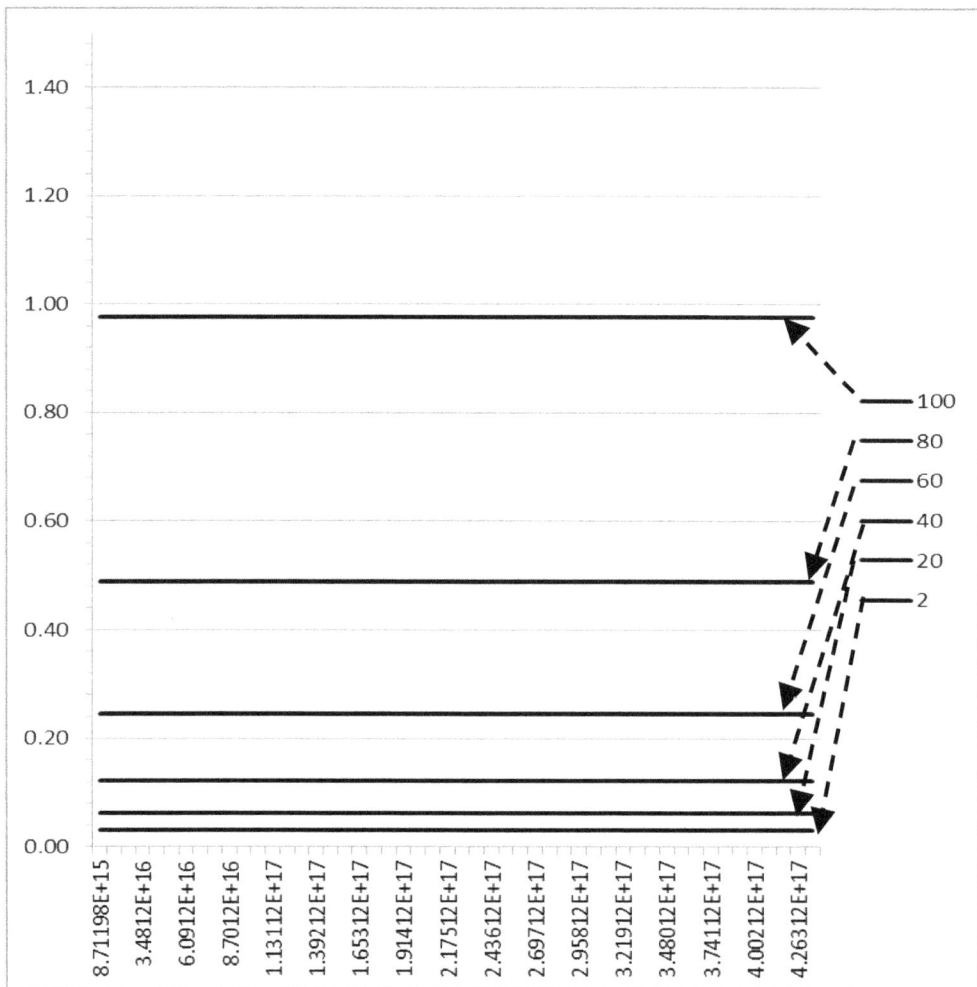

Figure 18.1 Plots of $\Omega_T(t)$ vs. time in seconds where Megaverse mass-energy density $\rho_{Mega} = \rho_{M0}\rho_{crit}$. The number and line ordering corresponds to the ordering in Table 18.1.

Note: If C_{T1} is very large than the Megaverse density $\boldsymbol{\rho_{M0}}$ factor will be very small. .

Note also, that the flatness of the curves in Fig. 18.1 suggests that we can use the average values given above as an approximation to $\Omega_T(t)$ in the Einstein equation for the various possible values of ρ_{M0}.

19. "Conventional" Universe Scale Factor for a Megaverse Influx Epoch

In this chapter we investigate the Influx-related scale factor $a_M(t)$ in eq. 17.19 for use in chapter 20 in the determination of the Hubble Parameter (Constant).

19.1 Influx-Driven Universe Expansion Epochs

In this section we summarize the epochs for the case of an influx-driven expansion under the assumption that it began at 380,000 years to facilitate comparison with astrophysical data. The influx, if any, that took place before that time is negligible in comparison with the influx after that time.

t	Time	Re a(t)	Im a(t)	a(t, ř)
t_{now}	13.8 Gyr	1	-6.38×10^{-93}	
Influx				$a_M(t)$
t_T	380,000 yr	$a(t_T)$		
Matter				$2H_0\Omega_m^{\frac{1}{2}}t)^{2/3}$
t_{RM}	2.2×10^{-5} Gyr	1.8×10^{-93}		
Radiation				$(2H_0\Omega_\gamma^{\frac{1}{2}}t)^{\frac{1}{2}}$
t_c	1.26×10^{-165} s	6.38×10^{-93}		
Big Bang metastate				Re $a_{BBRW}(t, ř) = (\gamma + a(t))/2$ Im $a_{BBRW}(t, ř) = (\gamma + a(t)) ř/2$
0	0 s	3.19×10^{-93}	-3.19×10^{-93}	

Figure 19.1. The Influx Epoch and Other Epochs.

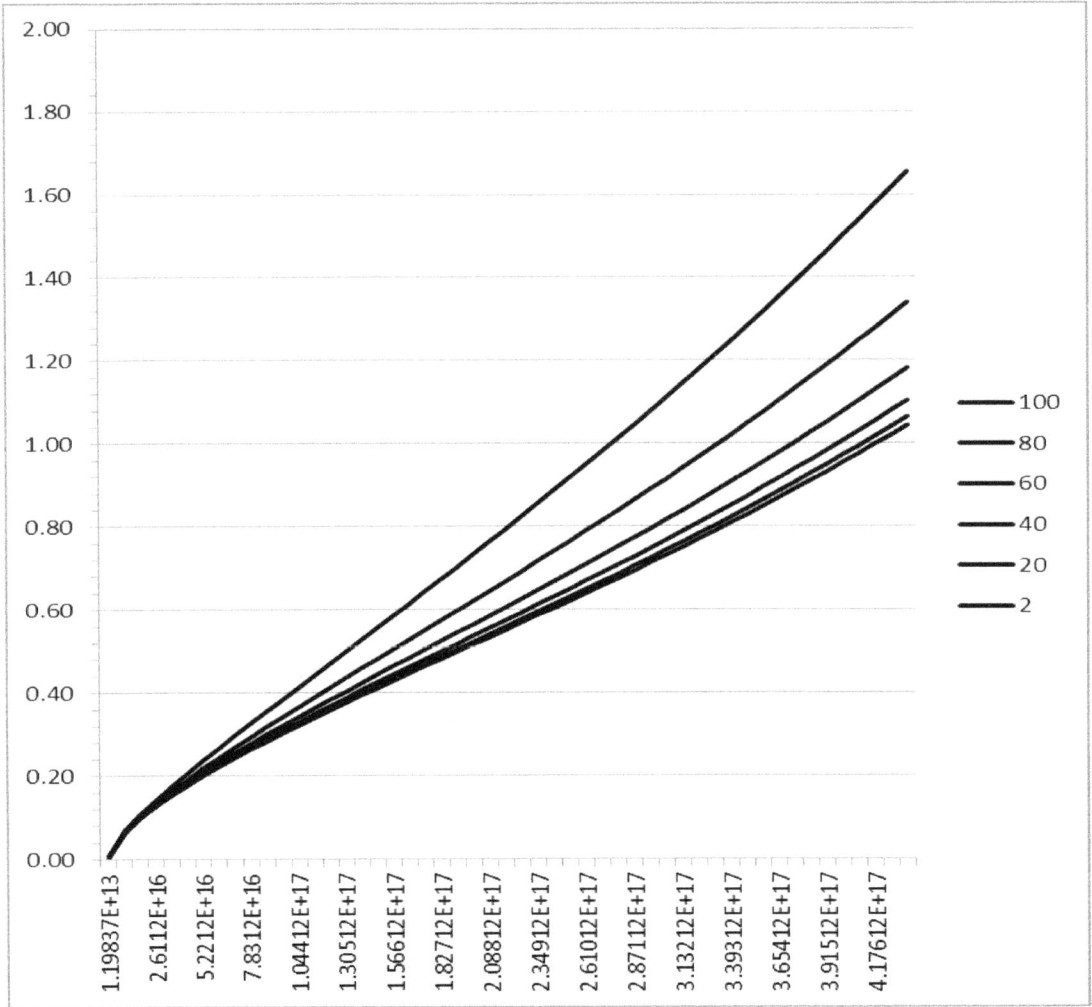

Figure 19.2. Scale factor $a_M(t)$ as a function of the time in seconds for six values of the Megaverse energy scale factor ρ_{M0}. The order of the lines is according to Table 18.1. The time ranges from 380,000 years to the present.

20. "Conventional" Hubble Constant Growth in Time

20.1 Influx Driven Hubble Constant in Time

The Einstein equation

$$\dot{a}^2(t) - 8\pi G\rho_{tot}a^2(t)/3 = -k \tag{11.19}$$

can be used to find the Hubble Parameter (Constant) as a function of time:

$$H(t) = (da/dt)/a(t) = [H_0^2\rho_{tot}(t)/\rho_{cri} - c^2k/a^2(t)]^{1/2} \tag{20.1}$$

where c is the speed of light and

$$\rho_{tot}(t) = \rho_{crit}(\Omega_H(t) + \Omega_T(t)) \tag{20.2}$$

$$\Omega_H(t) = \Omega_\gamma(t) + \Omega_m(t) + \Omega_d(t) + \Omega_\Lambda(t) \tag{9.10}$$

$$H_0^2 = 8\pi G\rho_{crit}/3 \tag{20.3}$$

Eq. 20.1 is plotted in Fig. 20.1. The negative values of H(t) squared (and the accompanying sharp edges in the curve) indicate the calculation does not capture the essence of the expansion properly.

For this reason we will examine a new approach to determining the expansion of the universe in the next chapter.

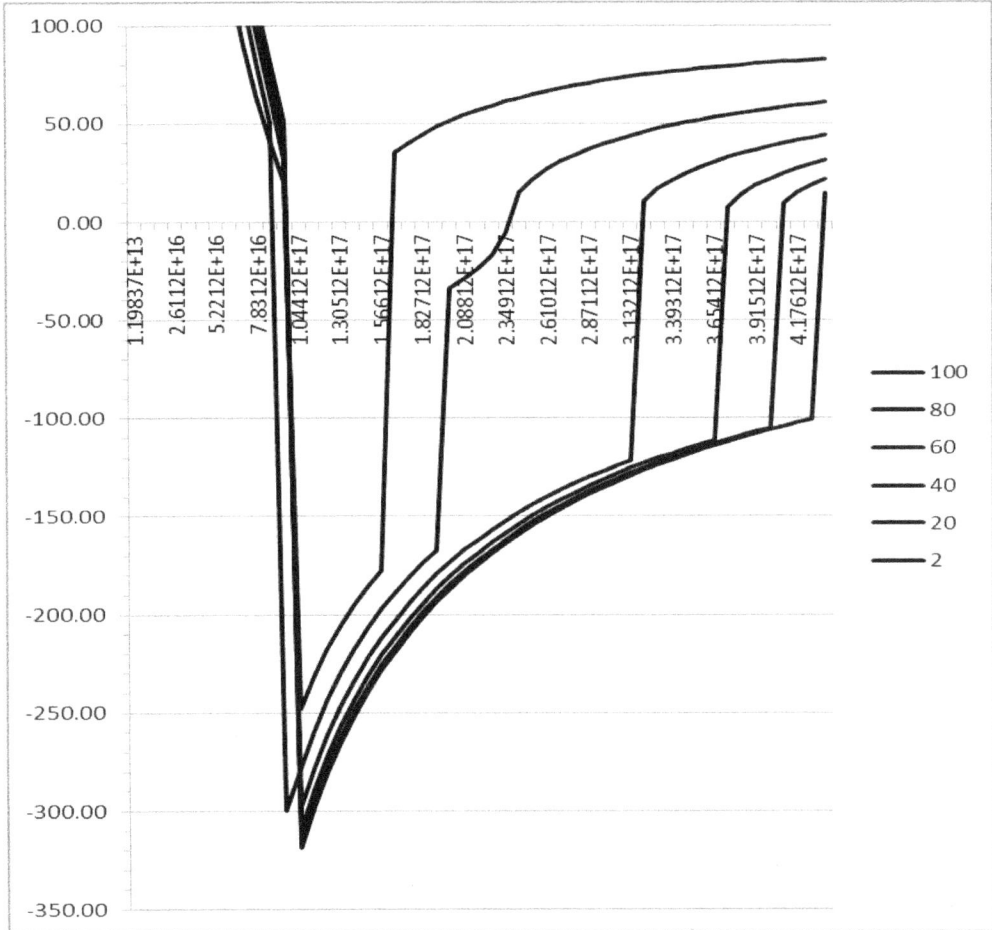

Figure 20.1. Plot of H(t) as a function of time in seconds. The apparently negative values of H(t) signal the fact that H(t) squared is negative at those points due to the dominance of the $c^2k/a^2(t)$ term in eq. 20.1.

Section 3: A NEW PHENOMENOLOGICAL THEORY OF THE HUBBLE PARAMETER

21. *"New"* Approach: Universe Expansion from the Big Bang Metastate

This chapter attempts to calculate the growth of the universe based on a phenomenological approximation to the growth of the universe from the Big Bang metastate to the present. It attempts to address the current impasse in our understanding of universe expansion. Something is wrong with our current understanding of the universe. A suggestion of this problem appears in the negative values of the Hubble Parameter squared in our calculation (chapter 20) which appears to be more than a result of our approximations.[92]

We begin with a look at the Hubble Parameter expansion experimental data. We then proceed to suggest a phenomenology that accounts for the data and nicely projects back to the Big Bang metastate discussed earlier in this book (and in Blaha (2004)).

21.1 Hubble Universe Expansion Experimental Data

There are a number of astrophysical studies of the universe that suggest that the Hubble Constant is *not* constant. Although there are significant margins of error it appears that the early universe "beginning" epoch at 380,000 years had a Hubble Constant of 67.8 km s^{-1} Mpc^{-1}.[93] More recently, red shift studies of quasars have given a Hubble Constant of 73.2 km s^{-1} Mpc^{-1}.[94] And studies of binary black hole merger gravity waves have given a Hubble Constant of 75.2 km s^{-1} Mpc^{-1} (and earlier of 78 km s^{-1} Mpc^{-1}). Another study of events at 1.8 billion ly yielded a Hubble Constant of 70.0

---

[92] The primary cause of the problem may be the use of the matter-dominated a(t) as the initial input for the calculation of the true a(t) taking account of a Megavers surface tension input term. However there appears to be more issues that lead to the problems with the Hubble Parameter calculation.

[93] See, for example, K. Aylor *et al*, arXiv:1811.00537v1 (2018) based on studies of the cosmological sound horizon.

[94] M. Soares-Santos *et al* , arXiv:1901.01540 (2019).

km s^{-1} Mpc^{-1}.[95] Further studies have given the Hubble Constants: 1. Of variable stars 73.2 km s^{-1} Mpc^{-1}, 2. Of light bent by distant galaxies 72.5 km s^{-1} Mpc^{-1}, and 3. Of Magellan Cepheids 74.03 ± 1.42 km s^{-1} Mpc^{-1}. [96]

The only apparent conclusion at this time is that there was a Hubble Parameter (Constant) H of approximately 67.8 km s^{-1} Mpc^{-1} early in the universe, and of perhaps 75.2 km s^{-1} Mpc^{-1} at the current time. Thus an increasing Hubble Constant.

For the purpose of discussing the apparent increase in H with time we average the above seven "recent" values of H in the spirit of Bayesian equal probability to obtain a recent time Hubble average of 73.7. Thus there appears to be a 9% increase in the Hubble Parameter over time.

21.2 A Phenomenological Fit to the Known Hubble Parameter Data

The radiation-dominated and the matter-dominated scale factors a(t) both are power laws in time as seen earlier in chapter 13. We therefore will assume that the true a(t) has a power law form:

$$a(t) = (t/t_{now})^{g + ht} \tag{21.1}$$

where g and h are constants. There is an "ht" term in the exponent based on the rise in H(t) noted above in the experimental data. The Hubble Parameter implied by eq. 21.1 is

$$H(t) = (da/dt)/a = g/t + h(1 + \ln(t/t_{now}) \tag{21.2}$$

If we set the value of H(t) at two values of time then g and h are determined. Based on the above discussion we use the experimental data:

$$H(t_c) \equiv H(380,000 \text{ yr}) = 67.8 \tag{21.3}$$
$$H(t_{now}) = 73.7$$

---

[95] B.P. Abbott *et al*, arXiv:1710.05835 (2017).
[96] J. T. Nielsen *et al*, Marginal evidence for cosmic acceleration from Type Ia supernovae, Nature Scientific Reports (2016); arXiv:1506.01354 (2015)

Eqs. 21.2 and 21.3 imply

$$h = (t_c H(t_c) - t_{now} H(t_{now}))[\ t_c - t_{now} + t_c \ln((t_c/t_{now})]^{-1} \qquad (21.5)$$
$$g = (H(t_{now}) - h)\ t_{now}$$

Substituting the parameter values of eq. 21.3 we obtain

$$h = 2.27402 \times 10^{-18} \qquad (21.6)$$
$$g = 0.000283993 = 2.83993 \times 10^{-4}$$

We use these parameters to calculate a(t), H(t), and Ω_T in the following plots at the end of the chapter for the time ranges: 1) from 380,000 years to the present, and 2) from the Big Bang metastate at $t = 1.1984 \times 10^{-165}$ s to the present.

The times are displayed in seconds and gigayears (Gyr0. In some plots a logarithmic plot was necessary because of the vast ranges of values. *We now turn to comments on the plots since some new and unforeseen features emerge.*

The calculation of Ω_T which may be in part pressure and in part Megaverse influx is based on the equation

$$\Omega_T = (H(t)^2 + c^2 k/a^2)/H_0^2 - \Omega_\gamma /a^4 - \Omega_M /a^3 - \Omega_\Lambda \qquad (21.7)$$

using

$$H(t) = (da/dt)/a(t) = [H_0^2 \rho_{tot}(t)/\rho_{cri} - c^2 k/a^2(t)]^{\frac{1}{2}}$$

21.3 Changing the Base Time from t_{now} to a different Base Time

The results above can be expressed in terms of a different time t_m

$$a'(t) = (t/t_m)^{g' + h't} \qquad (21.8)$$

using the same fixing of values with the Hubble Parameter data set to

$$H(t_1) = H'(t_1) \qquad (21.9)$$

$$H(t_2) = H'(t_2)$$

where

$$H'(t) = g'/t + h'(1 + \ln(t/t_m)) \tag{21.10}$$
$$t_1 = t_c$$
$$t_2 = t_{now}$$

The new parameters are

$$h' = \{[(g + ht_1)\ln(t_1/t_n)/\ln(t_1/t_m) - (g + ht_2)\ln(t_2/t_n)/\ln((t_2/t_m)]/(t_1 - t_2)$$
$$g' = (g + ht_2) \ln(t_2/t_n)/\ln((t_2/t_m) - h't_2 \tag{21.11}$$

Although the parameterization appears different the physical results are the same.[97]

21.4 Smooth H(t) Connection to Big Bang Metastate

The parameters in eqs. 21.1 and 21.2 were set by the H(t) data at 380,000 years and the present. If we extrapolate back to the end of the Big Bang metastate then we find a good match between the values of a(t) and H(t) of the Big Bang metastate and the extrapolation:

	H(t)
Big Bang Metastate	Big Bang Center 8.95×10^{217} km s^{-1} Mpc^{-1}
	Big Bang Edge $\quad 1.14 \times 10^{126}$ km s^{-1} Mpc^{-1}
Extrapolated H(t)	7.678×10^{180} km s^{-1} Mpc^{-1}

where the Big Bang values appear in section 13.6.2. Note that the extrapolated value is within the range of values in the Big Bang metastate and is roughly an average of the range of Big Bang values. Thus our H(t) fit (eqs. 21.1 and 21.2) extends smoothly back to the Big Bang.

The scale factor also shows a rough agreement. Using

---

[97] In that regard it is analogous to the renormalization group analysis of coupling constants with "coupling constants" $h = h(t_{now})$, $h' = h(t_m)$, and the times t_{now} and t_m being analogous to energy scales.

$$\text{Re } a_{BBRW}(t) \cong \gamma/2 + a(t)/2 \tag{13.3.2.17a}$$

where

$$\gamma/2 = 8.16 \times 10^{-93}$$

we see that

$$\text{Re } a_{BBRW}(t) \cong a(t)/2$$

giving us agreement in $a(t)$ within a factor of two.

Thus $H(t)$ and $a(t)$ are in a rather remarkable approximate agreement for the Big Bang metastate and our extrapolation at the cross over time 1.1984×10^{-165}s.

21.5 Fluctuating Behavior of a(t) and H(t) – The Big Dip

An examination of Figs. 21.1, 21.2 and 21.3 reveal a Big Dip in $H(t)$ which seems to have been unforeseen in astrophysical investigations. The Big Dip has a corresponding big peak in $\Omega_T(t)$ as shown in Fig. 21.5. The big peak occurs at approximately the same point in time as the Big Dip. Fig. 21.4 shows a corresponding minimum for $a(t)$.

The reason for these anomalies is uncertain although it is apparent that they are not a fault of the scale factor fit in eq. 21.1. A similar problem (negative H^2) appears in the calculations of chapters 17 through 20.

21.5.1 Location of the Big Dip in H(t)

The locations in time of the Big Dip events are:

Big Dip low point at $t = 8.71 \times 10^{13}$ s.
Ω_T plot peak at $t = 8.71 \times 10^{13}$ s

Since the times match it appears that the Big Dip is a result of massive growth in the mass-energy embodied in Ω_T.

Since the changeover from a radiation-dominated phase to a matter-dominated phase occurs at a slightly earlier time:

Radiation – Matter Domination Transition: $t = 1.48 \times 10^{12}$ s.

it seems reasonable to conclude the transition causes the Big Dip to occur. The matter dominated phase transition causes shrinkage as shown in a(t) in Fig. 21.4. The time delay between the transition and the Big Dip may be attributed to the time required for the transition to occur.

21 5.2 Overshoot in H(t)

The result of the Radiation-Matter transition is negative H(t) (and negative H(t) squared in chapter 20 for the surface tension energy influx model). H(t) "overshoots" and becomes negative. Crudely put, the clumping of matter in the matter-dominated phase could introduce a compactness that results in a decrease in universe size.

21.6 Mystery of the Big Dip in H(t) - A Scenario

At the Big Dip H(t) changes from a declining to a rising trajectory. Based on this fact and the Big_Bang-Megaverse_Driven model presented earlier the following scenario seems reasonable:

1. The initial peak, and immediate decline, in H(t) is due to the Y black body radiation phase pressure that decreases rapidly after the Big Bang metastate ends. Thus Ω_T declines rapidly (Fig. 21.7) with the Y pressure decline. (Note Ω_T is a sum of energy density and pressure.)

2.   The peak in Ω_T in Fig. 21.5 reflects an influx of energy that causes H(t) to begin increasing.

3. Ω_T continues to be significant and increasing as a(t) and H(t)rise to the present time.

4. In the future Ω_T will continue to rise. The energy increase that this situation implies suggests the part of our model where the surface tension force causes a significant influx of mass-energy from the Megaverse.

21.7 Current H(t() Increase Explained

The scenario presented in the previous section "explains" the rise of H(t) from 67.8 to an average 73.7.

21.8 An Interlude in the Eons

Based on the above analysis it appears that the universe is currently in an *interlude* following a decline in growth rate after the Big Bang, and the new beginning of major growth due to Megaverse energy influx that will last for eons.

21.9 Why is the Fit So Successful?

The phenomenological fit in eq. 21.1 appears to be a successful fit to the growth of the universe. It raises the question: Why is it so successful? It matches the Big Bang model at the beginning. It accounts for the observed rise in H(t). It shows an initial decline in H(t) after the end of the Big Bang epoch. It matches our expectation that the rise in H(t) is the result of an energy inflow from the Megaverse.

In this author's view its success is probably attributable to renormalization group self-similarity[98] as indicated in section 21.3 and its footnote. The evolution of universes may have laws of growth from a Big Bang metastate that are based on self-similar growth along the lines of the Renormalization group. This study is left to a future work when the growth of the universe is better understood experimentally.

---

[98] The self-similarity is based on the principle that the physics of the evolution of universes should be independent of the choice of the present time t_{now}. A change in t_{now} should result in a corresponding change in parameters—the essence of the Renormalization group. See section 21.3.

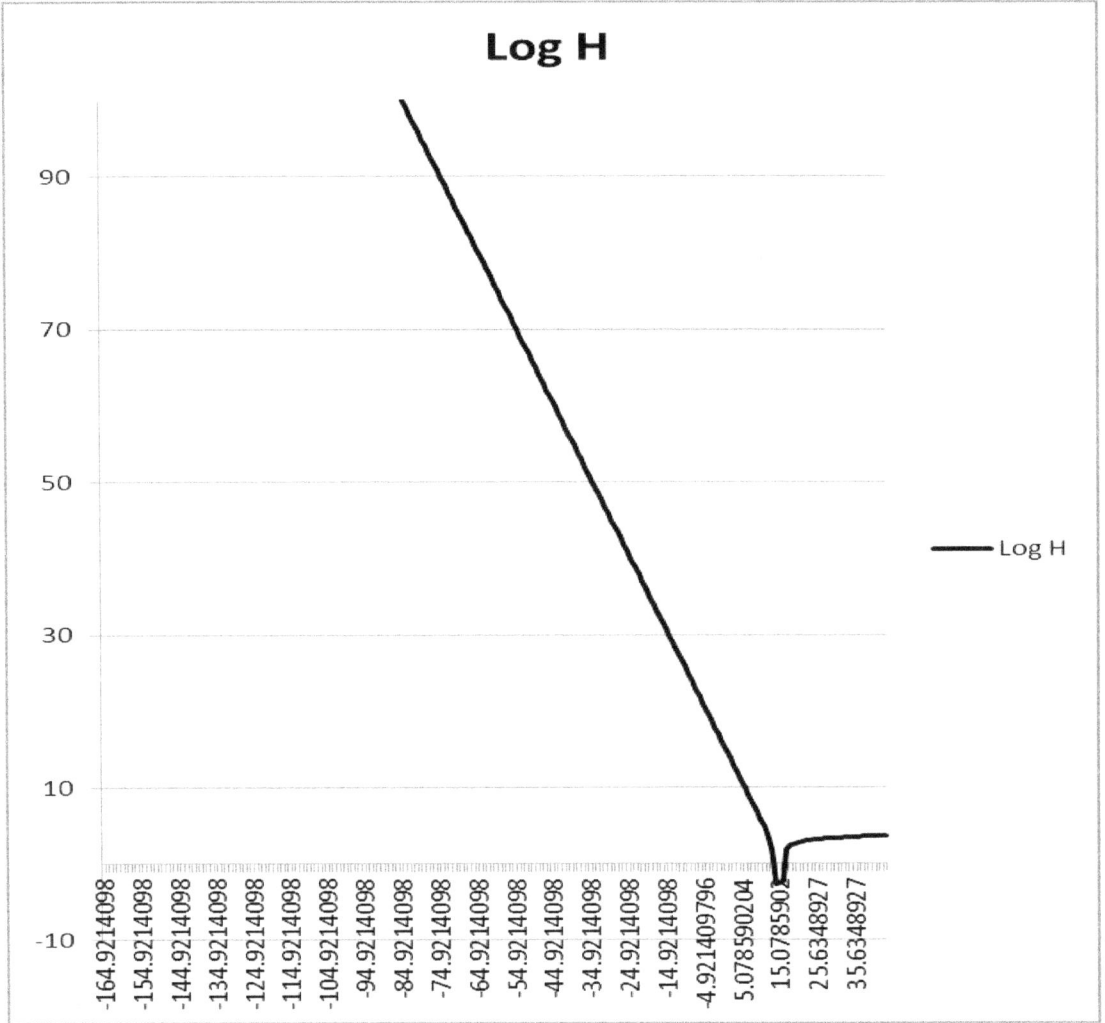

Figure 21.1. Log H(t) plotted vs. log t from the Big Bang metastate to the present. Note the dip below zero of H(t). Also note H(t) rises after the dip as measured experimentally.

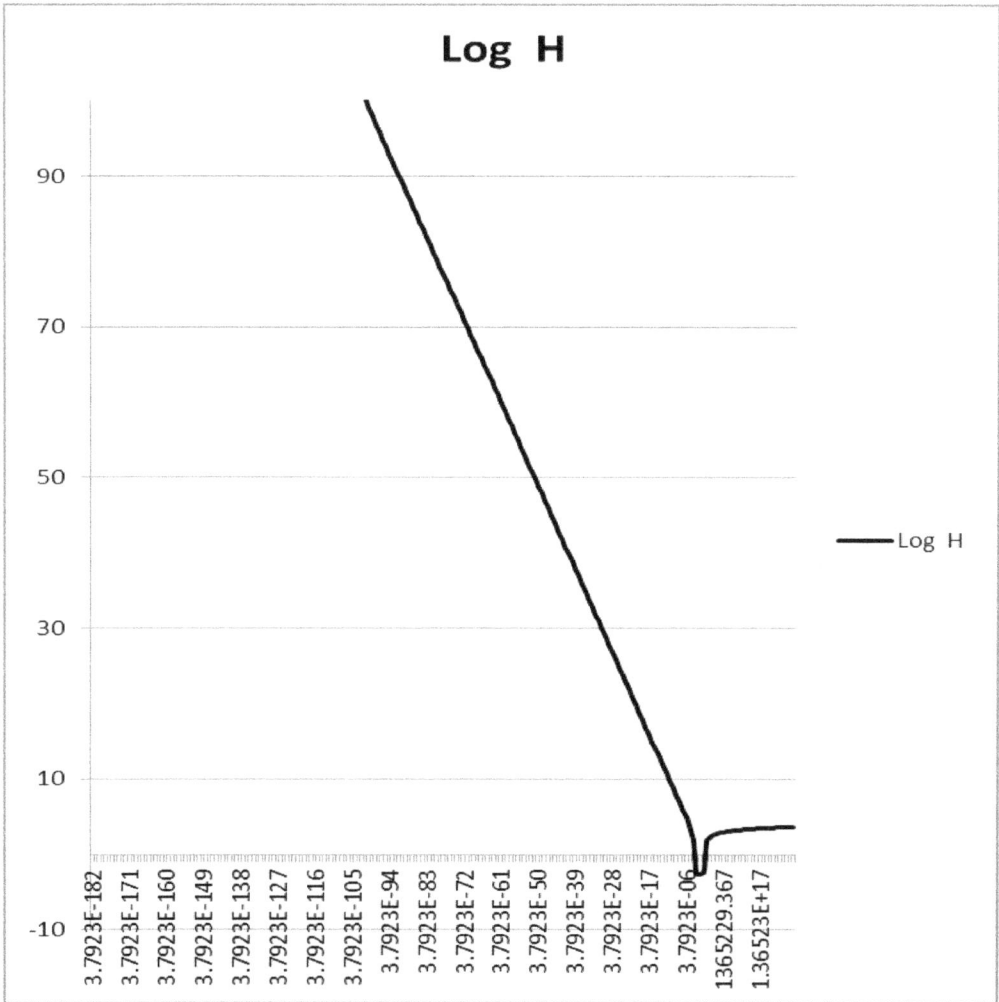

Figure 21.2. Log H(t) plotted vs. time in Gyr' from the Big Bang metastate to the present. Note the dip below zero of H(t). Also note H(t) rises after the dip as measured experimentally. The number following "E" is a power of ten.

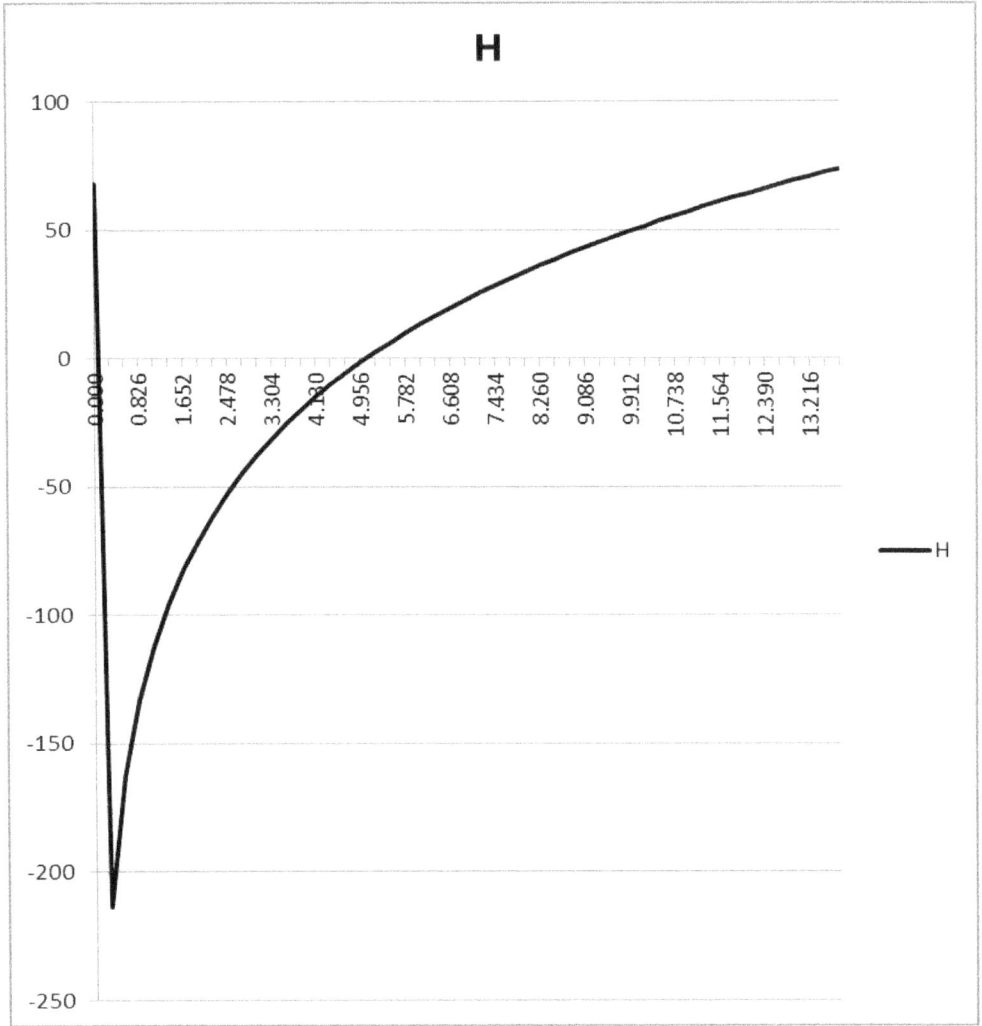

Figure 21.3. H(t) plotted vs. time in Gyr's from the year 380,000 to the present. Note the pronounced dip below zero of H(t). Also note H(t) rises after the dip as measured experimentally. The low point of the dip occurs at t = 8.71×10^{13} s.

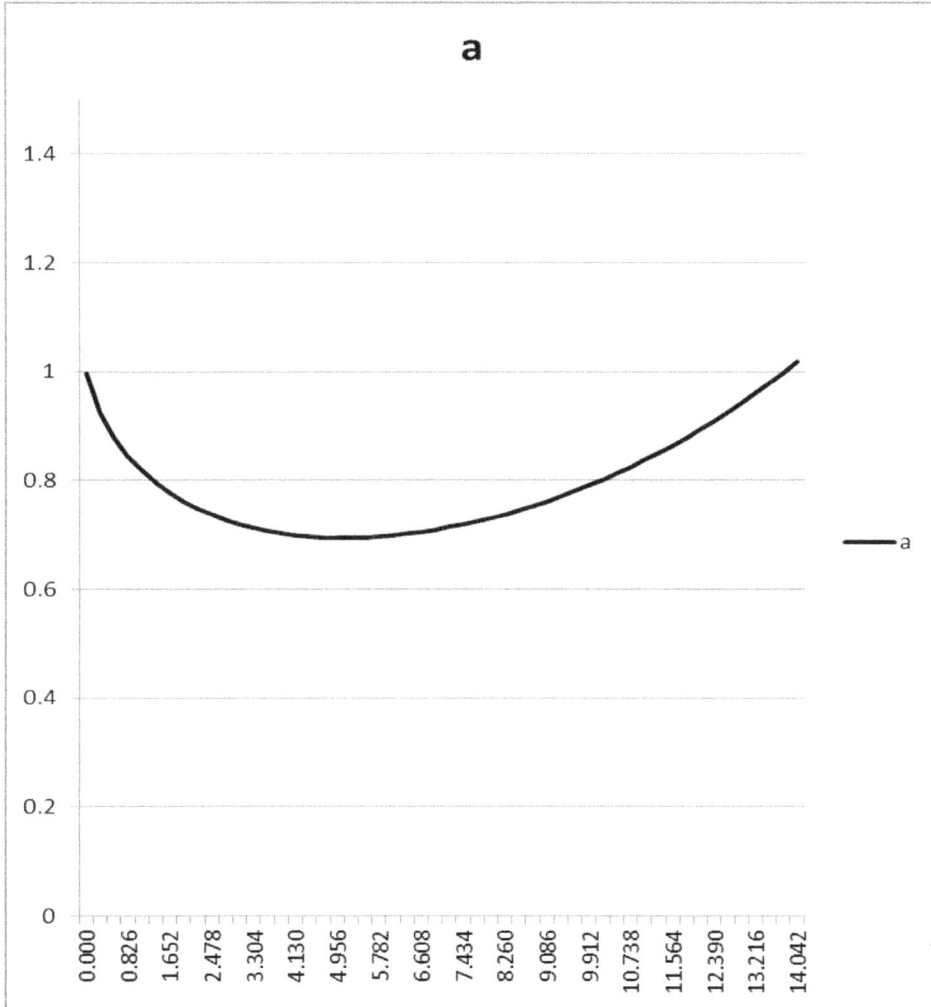

Figure 21.4. a(t) plotted vs. time in Gyr's from the year 380,000 to the present. Note the dip in a(t) (a smaller universe?) and then rises again to its current size.

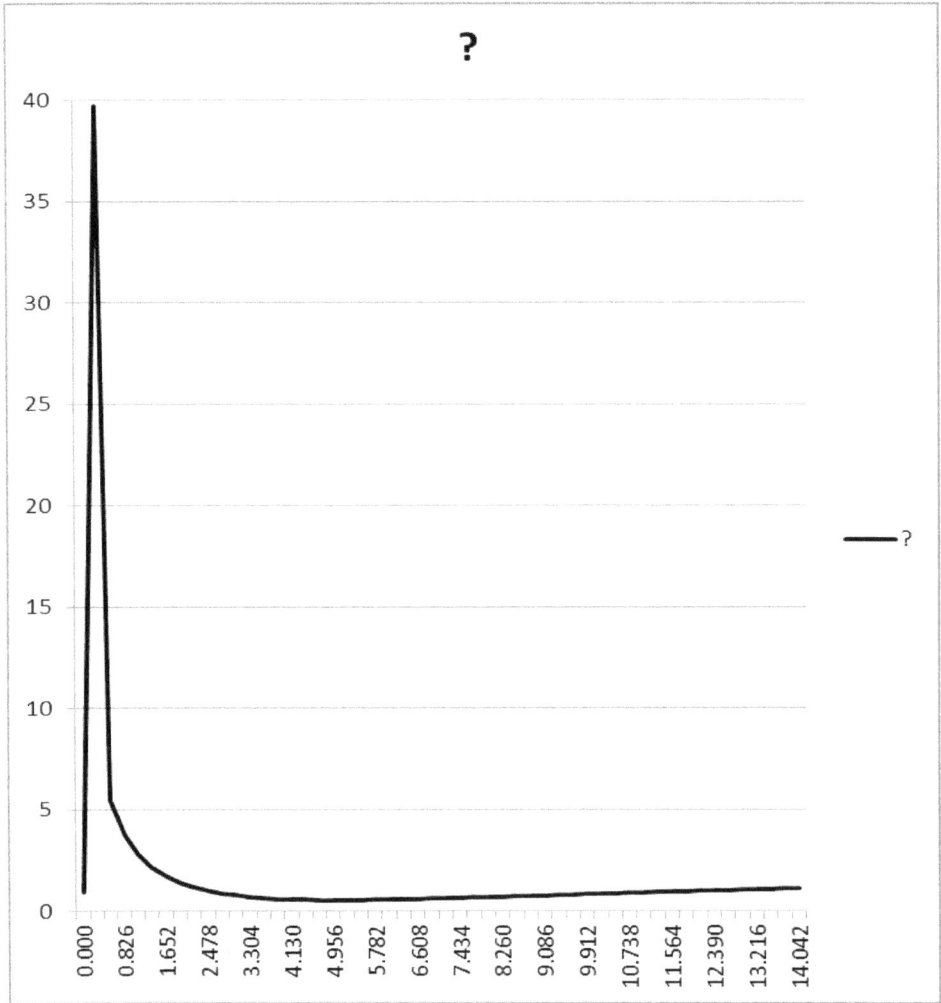

Figure 21.5. $\Omega_T(t)$ plotted vs. time in Gyr's from the year 380,000 to the present. Note the peak suggesting an influx into the universe at the transition to the matter-dominated phase, and then a decline followed by a raise.

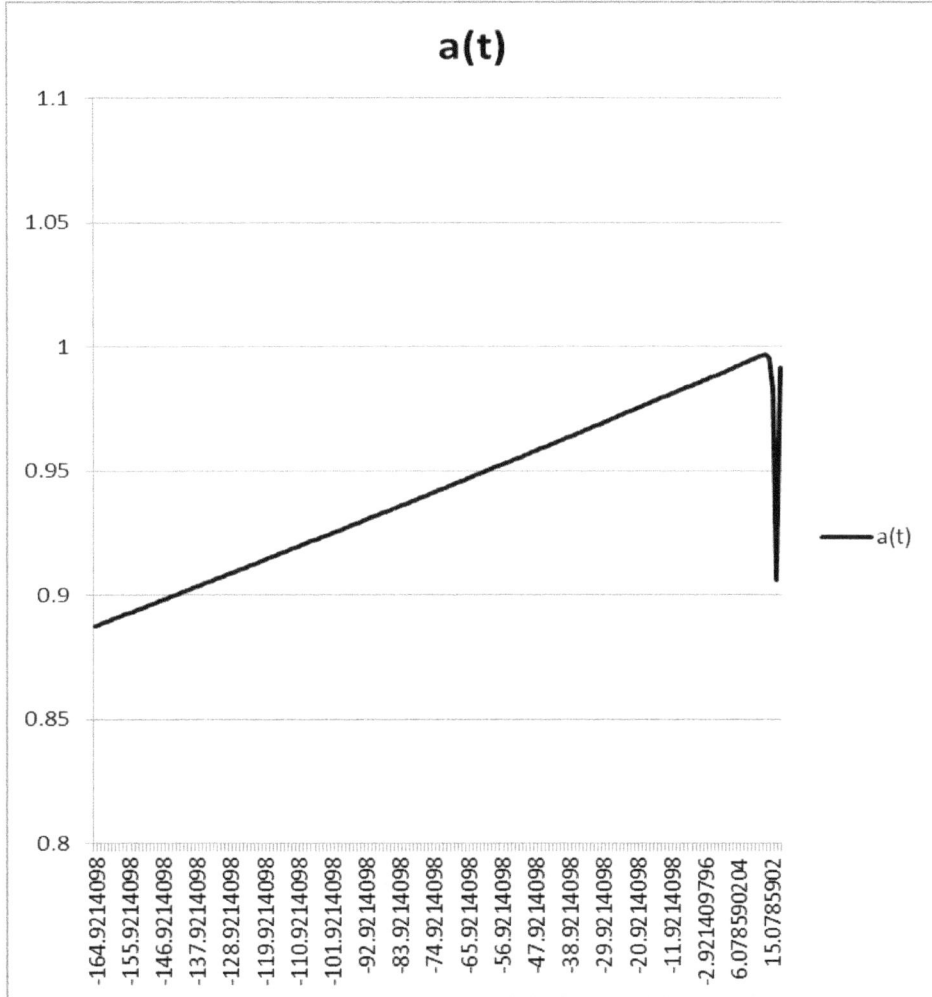

Figure 21.6. a(t) plotted vs. log time in seconds from the Big Bang metastate to to t = 13.65 Gyr. Note the dip in a(t) (a smaller universe?) after 380,000 yrs followed by a rise that we are now experiencing and will continue.

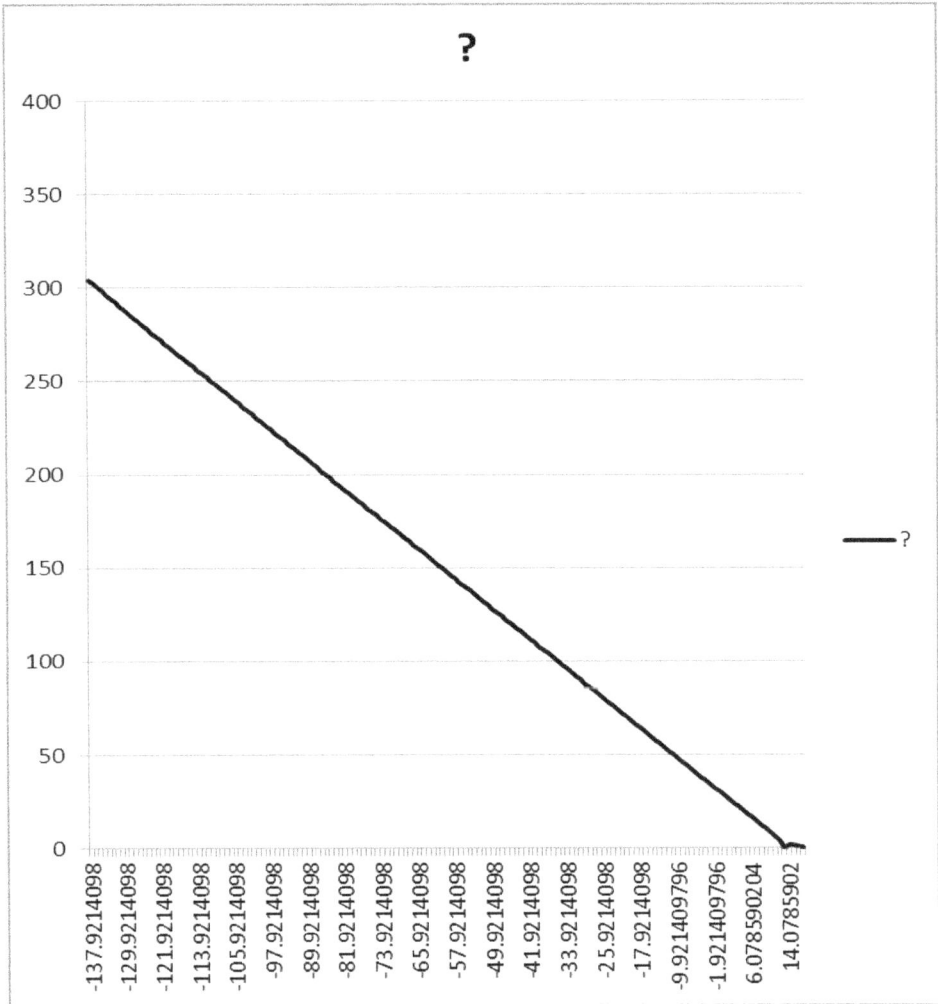

Figure 21.7. Log $\Omega_T(t)$ plotted vs. log time in seconds from 3.8×10^{-155} Gyr to the 13.65 Gyr. Note the scarcely visible fluctuation in dip in $\Omega_T(t)$ (a smaller universe?) after 380,000 yrs. See Fig. 21.5 for a detailed view.

22. Evidence for the Megaverse

Earlier (chapter 1) we gave evidence for the possible existence of a Megaverse containing (perhaps) many, many universes. In the preceding chapters we have developed the part of our Hybrid Quantum Big_Bang-Megaverse_Driven Universe that has an influx of mass-energy from the Megaverse fueling the increasing rate of expansion of our universe. In this chapter we list the theoretical and experimental support for the Megaverse.

The increasing value of the Hubble Constant is perhaps the strongest support for a Megaverse since it implies the mass-energy of our universe is increasing. A Megaverse influx is the most reasonable explanation for the increasing Hubble Constant.

The circumstantial evidence given in chapter 1 further supports the existence of the Megaverse.

22.1 Theoretical Evidence for Other Universes and External Matter

The circumstantial (and substantial) theoretical evidence, which have been subjects of discussion for some time, are:

1. The need for an external 'clock' to measure 'time' knowing that it is to some extent relative and local.

2. The need for a 'quantum observer' to complete the understanding of quantum gravity as described by the Wheeler-DeWitt equation and other efforts to develop a quantum gravity.

3. The need for other universes to provide theoretical measuring platforms for quantities beyond the charge and mass-energy of the universe. We think here of

the other quantum numbers of particles and particle number operators such as Baryon number.

4. The need for an ultimate source of mass and inertia in our universe.

22.2 Possible Experimental Evidence for the Megaverse

At first glance it would seem impossible to produce evidence for the existence of other universes. However there are subtle means by which we can 'sense' experimentally 'nearby' universes should they exist. The mechanism would appear to be gravitational effects exerted on objects within our universe by unseen nearby objects of enormous mass. Currently there appears to be three experimental suggestions for the existence of 'nearby' universes:

1. The existence of Great Attractors. (See section 1.2.1)

2. The existence of Bright Bumps in Universe Suggesting Collision with Another Universe.

3. The existence of a Cold Spot in Universe Suggesting a Collision with Another Universe.

4. Hubble Rate Increase

Appendix A. Some Physical Constants Used in Calculating Numerical Expressions

Some physical constants that we found to be of use in the evaluation of expressions are (assuming units with $c = \hbar = 1$):

$$G \equiv 7.39 \times 10^{-29} \text{ gm}^{-1} \text{ cm} \qquad \text{(A.1a)}$$
$$G \equiv 2.6 \times 10^{-66} \text{ cm}^2 \qquad \text{(A.1b)}$$
$$G \equiv 2.91 \times 10^{-87} \text{ s}^2 \qquad \text{(A.1c)}$$

$$H_0 \equiv 2.133 \times 10^{-33} h \text{ ev} \qquad \text{(A.2a)}$$
$$H_0 \equiv 1.08 \times 10^{-28} h \text{ cm}^{-1} \qquad \text{(A.2b)}$$
$$H_0 \equiv 3.24 \times 10^{-18} h \text{ s}^{-1} \qquad \text{(A.2c)}$$
$$h = .689 \qquad \text{(A.2d)}$$
$$H_0 = 100 \, h \text{ km s}^{-1} \text{ Mpc}^{-1} \qquad \text{(A.2e)}$$
$$1 \text{ s}^{-1} = 3.0864 \times 10^{-20} \text{ km s}^{-1} \text{ Mpc}^{-1} \qquad \text{(A.2f)}$$

$$GH_0 \equiv 7.98 \times 10^{-57} h \text{ gm}^{-1} \qquad \text{(A.3)}$$

$$\rho_{\text{crit}} \equiv 1.88 \times 10^{-29} h^2 \text{ gm cm}^{-3} \qquad \text{(A.4)}$$

$$M_{\text{Planck}} \equiv 1.22 \times 10^{28} \text{ ev} \qquad \text{(A.5a)}$$
$$M_{\text{Planck}} \equiv 2.18 \times 10^{-5} \text{ g} \qquad \text{(A.5b)}$$

$$M_{\text{Planck}} \equiv 6.20 \times 10^{32} \text{ cm}^{-1} \qquad \text{(A.6a)}$$
$$\text{Planck Length} = M_{\text{Planck}}^{-1} \equiv 1.61 \times 10^{-33} \text{ cm} \qquad \text{(A.6b)}$$

$$M_{Planck} \equiv 1.85 \times 10^{43} \text{ s}^{-1} \tag{A.7a}$$

$$\text{Planck time} = M_{Planck}^{-1} \equiv 5.41 \times 10^{-44} \text{ s} \tag{A.7b}$$

$$1 \text{ eV} \equiv 5.08 \times 10^{4} \text{ cm}^{-1} \tag{A.8a}$$

$$1 \text{ eV} \equiv 1.52 \times 10^{15} \text{ s}^{-1} \tag{A.8b}$$

$$1 \text{ eV}/c^2 \equiv 1.783 \times 10^{-33} \text{ g} \tag{A.8c}$$

$$1 \text{ g} \equiv 2.85 \times 10^{37} \text{ cm}^{-1} \tag{A.8d}$$

$$\kappa \equiv 4.38 \text{ °K}^{-1} \text{ cm}^{-1} \tag{A.9}$$

$$\kappa \equiv 1.31 \times 10^{11} \text{ °K}^{-1} \text{ s}^{-1} \tag{A.10}$$

$$\kappa \equiv 8.62 \times 10^{-5} \text{ ev °K}^{-1} \tag{A.11}$$

$$1 \text{ Gyr} = 3.16 \times 10^{16} \text{ s} \tag{A.12}$$

where κ is Boltzmann's constant.

Appendix B. Defined Quantities in the Text

This appendix lists constants defined by the Particle Data Group and in the text.

Constants based on M. Tanabashi *et al op. cit.*

h = Hubble parameter = 0.678(9)

ρ_{crit} = Critical density = $1.87840(9)\ h^2 \times 10^{-29}$ g/cm^{-3}

Ω_Λ = Dark Energy density/$\rho_{cr} = \rho_{de}/\rho_{cr} = 0.692 \pm 0.012$

$\Omega_d = \Omega_c$ = Cold Dark matter density/$\rho_{cr} = \rho_c/\rho_{cr} = 0.1186(20)\ h^{-2}$

Ω_b = Baryon density/$\rho_{cr} = \rho_b/\rho_{cr} = 0.02226\ h^{-2}$

$\Omega_m = \Omega_b + \Omega_c$

$\quad$ = pressureless Matter density/$\rho_{cr} = \rho_m/\rho_{cr} = 0.308 \pm 0.012$

$\Omega_{mtot} = \Omega_m + \Omega_\Lambda = 1.000 \pm 0.024$

$t_0 = t_{now}$ = Age of universe = 13.80 ± 0.04 Gyr

$r_{universe}(t_{now})$ = visible radius of the universe = 4.314×10^{28} cm

$$(13.1.3)$$

Ω_γ = radiation density/$\rho_{cr} = \rho_\gamma/\rho_{cr} = 2.473h^{-2} \times 10^{-5}\ (T/2.7255)^4 h^{-2}$

$\quad = 5.38 \times 10^{-5}$

$$(13.1.4)$$

$k = 5.56 \times 10^{-57}$ cm^{-2}

Constants defined in the text:

$$C_M = C_T G \rho_{Mega} \tag{9.8}$$

Robertson-Walker volume:

$$V = 2\pi^2 R^3 \tag{10.1.2b}$$

$$\alpha = \gamma/\delta = 1 \tag{10.6.3}$$

From the current CMB temperature ($T_0 = 2.7255$ °K) we find

$$\kappa T = \kappa T_0/a(t_{now}) = \kappa T_0 = 2.35 \times 10^{-4} \text{ ev} \qquad (13.3.1.1)$$
$$x = \pi^{3/2}\kappa T_0/(2M_c) \cong 5.3 \times 10^{-32} \qquad (13.3.1.2)$$

$$H_0 = 100 \text{ h km s}^{-1} \text{ Mpc}^{-1} = h \times (9.777752 \text{ Gyr})^{-1} =$$
$$= 1.1 \times 10^{-28} \text{h cm}^{-1} \equiv 3.24 \times 10^{-20} \text{h s}^{-1} \qquad (13.2.1.2b)$$
$$= 5.56 \times 10^{-57} \text{ cm}^{-2} \qquad (13.2.1.2c)$$

$$\Omega_m + \Omega_\Lambda \cong 1.00$$

$$\xi = k/H_0^2 \cong 1 \qquad (13.2.1.3)$$

$$a(t_{now}) = 1 \qquad (13.2.2.4)$$

$$H_0\Omega_\Lambda^{1/2} = 1.83 \times 10^{-18} \text{ s}^{-1} \qquad (13.2.2.6)$$

$$t_E = t_{now} + (H_0\Omega_\Lambda^{1/2})^{-1}\ln(.76) = 2.87 \times 10^{17} \text{ s} \qquad (13.2.2.7)$$
$$t_{now} = 4.35 \times 10^{17} \text{ s}$$
$$t_{RM} = 7 \times 10^{11} \text{ s} = 2.22 \times 10^{-5} \text{ Gyr} \qquad (13.2.3.12)$$
$$t_c = 1.26 \times 10^{-165} \text{ s} \cong 10^{-165} \text{ s} \qquad (13.3.2.15)$$
$$t_T = 380,000 \text{ years}$$
$$= 1.19837 \times 10^{13} \text{ sec}$$
$$= 1.205 \times 10^{-11} \text{ Gyr}$$

$$\kappa T = \kappa T_0/a(t_{now}) = \kappa T_0 = 2.35 \times 10^{-4} \text{ ev} \qquad (13.3.1.1)$$

$$\varpi = k^{-1/2}M_c = 8.31 \times 10^{60} \qquad (13.3.2.2)$$

$$\gamma = x/\varpi = 6.38 \times 10^{-93} \qquad (13.3.2.9)$$

REFERENCES

Akhiezer, N. I., Frink, A. H. (tr), 1962, *The Calculus of Variations* (Blaisdell Publishing, New York, 1962).

Bjorken, J. D., Drell, S. D., 1964, *Relativistic Quantum Mechanics* (McGraw-Hill, New York, 1965).

Bjorken, J. D., Drell, S. D., 1965, *Relativistic Quantum Fields* (McGraw-Hill, New York, 1965).

Blaha, S., 1998, *Cosmos and Consciousness* (Pingree-Hill Publishing, Auburn, NH, 1998).

_____, 2002, *A Finite Unified Quantum Field Theory of the Elementary Particle Standard Model and Quantum Gravity Based on New Quantum Dimensions™ & a New Paradigm in the Calculus of Variations* (Pingree-Hill Publishing, Auburn, NH, 2002).

_____, 2003, *A Finite Unified Quantum Field Theory of the Elementary Particle Standard Model and Quantum Gravity Based on New Quantum Dimensions™ and a New Paradigm in the Calculus of Variations* (Pingree-Hill Publishing, Auburn, NH, 2003).

_____, 2004, *Quantum Big Bang Cosmology: Complex Space-time General Relativity, Quantum Coordinates™Dodecahedral Universe, Inflation, and New Spin 0, ½, 1 & 2 Tachyons & Imagyons* (Pingree-Hill Publishing, Auburn, NH, 2004).

_____, 2005a, *Quantum Theory of the Third Kind: A New Type of Divergence-free Quantum Field Theory Supporting a Unified Standard Model of Elementary Particles and Quantum Gravity based on a New Method in the Calculus of Variations* (Pingree-Hill Publishing, Auburn, NH, 2005).

_____, 2005b, *The Metatheory of Physics Theories, and the Theory of Everything as a Quantum Computer Language* (Pingree-Hill Publishing, Auburn, NH, 2005).

_____, 2005c, *The Equivalence of Elementary Particle Theories and Computer Languages: Quantum Computers, Turing Machines, Standard Model, Superstring Theory, and a Proof that Gödel's Theorem Implies Nature Must Be Quantum* (Pingree-Hill Publishing, Auburn, NH, 2005).

_____, 2006a, *The Foundation of the Forces of Nature* (Pingree-Hill Publishing, Auburn, NH, 2006).

_____, 2006b, *A Derivation of ElectroWeak Theory based on an Extension of Special Relativity; Black Hole Tachyons; & Tachyons of Any Spin.* (Pingree-Hill Publishing, Auburn, NH, 2006).

_____, 2007a, *Physics Beyond the Light Barrier: The Source of Parity Violation, Tachyons, and A Derivation of Standard Model Features* (Pingree-Hill Publishing, Auburn, NH, 2007).

_____, 2007b, *The Origin of the Standard Model: The Genesis of Four Quark and Lepton Species, Parity Violation, the ElectroWeak Sector, Color SU(3), Three Visible Generations of Fermions, and One Generation of Dark Matter with Dark Energy* (Pingree-Hill Publishing, Auburn, NH, 2007).

_____, *2008a, A Direct Derivation of the Form of the Standard Model From GL(16) (Pingree-Hill Publishing, Auburn, NH, 2008).*

_____, 2008b, *A Complete Derivation of the Form of the Standard Model With a New Method to Generate Particle Masses Second Edition* (Pingree-Hill Publishing, Auburn, NH, 2008)

_____, 2009, *The Algebra of Thought & Reality: The Mathematical Basis for Plato's Theory of Ideas, and Reality Extended to Include A Priori Observers and Space-Time Second Edition* (Pingree-Hill Publishing, Auburn, NH, 2009).

_____, 2010a, *Operator Metaphysics: A New Metaphysics Based on a New Operator Logic and a New Quantum Operator Logic that Lead to a Mathematical Basis for Plato's Theory of Ideas and Reality* (Pingree-Hill Publishing, Auburn, NH, 2010).

_____, 2010b, *The Standard Model's Form Derived from Operator Logic, Superluminal Transformations and GL(16)* (Pingree-Hill Publishing, Auburn, NH, 2010).

_____, 2010c, *SuperCivilizations: Civilizations as Superorganisms* (McMann-Fisher Publishing, Auburn, NH, 2010).

_____, 2011a, *21*[st] *Century Natural Philosophy Of Ultimate Physical Reality* (McMann-Fisher Publishing, Auburn, NH, 2011).

_____, 2011b, *All the Universe! Faster Than Light Tachyon Quark Starships & Particle Accelerators with the LHC as a Prototype Starship Drive Scientific Edition* (Pingree-Hill Publishing, Auburn, NH, 2011).

_____, 2011c, *From Asynchronous Logic to The Standard Model to Superflight to the Stars* (Blaha Research, Auburn, NH, 2011).

_____, 2012a, *From Asynchronous Logic to The Standard Model to Superflight to the Stars volume 2: Superluminal CP and CPT, U(4) Complex General Relativity and The Standard Model, Complex Vierbein General Relativity, Kinetic Theory, Thermodynamics* (Blaha Research, Auburn, NH, 2012).

_____, 2012b, *Standard Model Symmetries, And Four And Sixteen Dimension Complex Relativity; The Origin Of Higgs Mass Terms* (Blaha Reasearch, Auburn, NH, 2012).

_____, 2013a, *Multi-Stage Space Guns, Micro-Pulse Nuclear Rockets, and Faster-Than-Light Quark-Gluon Ion Drive Starships* (Blaha Research, Auburn, NH, 2013).

_____, 2013b, *The Bridge to Dark Matter; A New Sister Universe; Dark Energy; Inflatons; Quantum Big Bang; Superluminal Physics; An Extended Standard Model Based on Geometry* (Blaha Reasearch, Auburn, NH, 2013).

_____, 2014a, *Universes and Megaverses: From a New Standard Model to a Physical Megaverse; The Big Bang; Our Sister Universe's Wormhole; Origin of the Cosmological Constant, Spatial Asymmetry of the Universe, and its Web of Galaxies; A Baryonic Field between Universes and Particles; Megaverse Extended Wheeler-DeWitt Equation* (Blaha Reasearch, Auburn, NH, 2014).

_____, 2014b, *All the Megaverse! Starships Exploring the Endless Universes of the Cosmos Using the Baryonic Force* (Blaha Research, Auburn, NH, 2014).

_____, 2014c, *All the Megaverse! II Between Megaverse Universes: Quantum Entanglement Explained by the Megaverse Coherent Baryonic Radiation Devices – PHASERs Neutron Star Megaverse Slingshot Dynamics Spiritual and UFO Events, and the Megaverse Microscopic Entry into the Megaverse* (Blaha Research, Auburn, NH, 2014).

_____, 2015a, *PHYSICS IS LOGIC PAINTED ON THE VOID: Origin of Bare Masses and The Standard Model in Logic, U(4) Origin of the Generations, Normal and Dark Baryonic Forces, Dark Matter, Dark Energy, The Big Bang, Complex General Relativity, A Megaverse of Universe Particles* (Blaha Research, Auburn, NH, 2015).

_____, 2015b, *PHYSICS IS LOGIC Part II: The Theory of Everything, The Megaverse Theory of Everything, U(4)⊗U(4) Grand Unified Theory (GUT), Inertial Mass = Gravitational Mass, Unified Extended Standard Model and a New Complex General Relativity with Higgs Particles, Generation Group Higgs Particles* (Blaha Research, Auburn, NH, 2015).

_____, 2015c, *The Origin of Higgs ("God") Particles and the Higgs Mechanism: Physics is Logic III, Beyond Higgs – A Revamped Theory With a Local Arrow of Time, The Theory of Everything Enhanced, Why Inertial Frames are Special, Universes of the Mind* (Blaha Research, Auburn, NH, 2015).

_____, 2015d, *The Origin of the Eight Coupling Constants of The Theory of Everything: U(8) Grand Unified Theory of Everything (GUTE), S^8 Coupling Constant Symmetry, Space-Time Dependent Coupling Constants, Big Bang Vacuum Coupling Constants, Physics is Logic IV* (Blaha Research, Auburn, NH, 2015).

_____, 2016a, *New Types of Dark Matter, Big Bang Equipartition, and A New U(4) Symmetry in the Theory of Everything: Equipartition Principle for Fermions, Matter is 83.33% Dark,*

Penetrating the Veil of the Big Bang, Explicit QFT Quark Confinement and Charmonium, Physics is Logic V (Blaha Research, Auburn, NH, 2016).

_____, 2016b, *The Periodic Table of the 192 Quarks and Leptons in The Theory of Everything: The U(4) Layer Group, Physics is Logic VI* (Blaha Research, Auburn, NH, 2016).

_____, 2016c, *New Boson Quantum Field Theory, Dark Matter Dynamics, Dark Matter Fermion Layer Mixing, Genesis of Higgs Particles, New Layer Higgs Masses, Higgs Coupling Constants, Non-Abelian Higgs Gauge Fields, Physics is Logic VII* (Blaha Research, Auburn, NH, 2016).

_____, 2016d, *Unification of the Strong Interactions and Gravitation: Quark Confinement Linked to Modified Short-Distance Gravity; Physics is Logic VIII* (Blaha Research, Auburn, NH, 2016).

_____, 2016e, *MoND: Unification of the Strong Interactions and Gravitation II, Quark Confinement Linked to Large-Scale Gravity, Physics is Logic IX* (Blaha Research, Auburn, NH, 2016).

_____, 2016f, *CQ Mechanics: A Unification of Quantum & Classical Mechanics, Quantum/Semi-Classical Entanglement, Quantum/Classical Path Integrals, Quantum/Classical Chaos* (Blaha Research, Auburn, NH, 2016).

_____, 2016g, *GEMS: Unified Gravity, ElectroMagnetic and Strong Interactions: Manifest Quark Confinement, A Solution for the Proton Spin Puzzle, Modified Gravity on the Galactic Scale* (Pingree Hill Publishing, Auburn, NH, 2016).

_____, 2016h, *Unification of the Seven Boson Interactions based on the Riemann-Christoffel Curvature Tensor* (Pingree Hill Publishing, Auburn, NH, 2016).

_____, 2017a, *Unification of the Eleven Boson Interactions based on 'Rotations of Interactions'* (Pingree Hill Publishing, Auburn, NH, 2017).

_____, 2017b, *The Origin of Fermions and Bosons, and Their Unification* (Pingree Hill Publishing, Auburn, NH, 2017).

_____, 2017c, *Megaverse: The Universe of Universes* (Pingree Hill Publishing, Auburn, NH, 2017).

_____, 2017d, *SuperSymmetry and the Unified SuperStandard Model* (Pingree Hill Publishing, Auburn, NH, 2017).

_____, 2017e, *From Qubits to the Unified SuperStandard Model with Embedded SuperStrings: A Derivation* (Pingree Hill Publishing, Auburn, NH, 2017).

_____, 2017f, *The Unified SuperStandard Model in Our Universe and the Megaverse: Quarks, ... ,* (Pingree Hill Publishing, Auburn, NH, 2017).

_____, 2018a, *The Unified SuperStandard Model and the Megaverse SECOND EDITION A Deeper Theory based on a New Particle Functional Space that Explicates Quantum Entanglement Spookiness (Volume 1)* (Pingree Hill Publishing, Auburn, NH, 2018).

_____, 2018b, *Cosmos Creation: The Unified SuperStandard Model, Volume 2, SECOND EDITION* (Pingree Hill Publishing, Auburn, NH, 2018).

_____, 2018c, *God Theory* (Pingree Hill Publishing, Auburn, NH, 2018).

_____, 2018d, *Immortal Eye: God Theory: Second Edition* (Pingree Hill Publishing, Auburn, NH, 2018).

_____, 2018e, *Unification of God Theory and Unified SuperStandard Model THIRD EDITION* (Pingree Hill Publishing, Auburn, NH, 2018).

_____, 2019a, *Calculation of: QED $\alpha = 1/137$, and Other Coupling Constants of the Unified SuperStandard Theory* (Pingree Hill Publishing, Auburn, NH, 2019).

_____, 2019b, *Coupling Constants of the Unified SuperStandard Theory SECOND EDITION: We Find the Fine Structure Constant 1/137.0359801, and so: OUR UNIVERSE AND LIFE! Also a Universal Eigenvalue Function for all Known Interactions, And Running Coupling Constants to all Perturbative Orders* (Pingree Hill Publishing, Auburn, NH, 2019).

Eddington, A. S., 1952, *The Mathematical Theory of Relativity* (Cambridge University Press, Cambridge, U.K., 1952).

Fant, Karl M., 2005, *Logically Determined Design: Clockless System Design With NULL Convention Logic* (John Wiley and Sons, Hoboken, NJ, 2005).

Feinberg, G. and Shapiro, R., 1980, *Life Beyond Earth: The Intelligent Earthlings Guide to Life in the Universe* (William Morrow and Company, New York, 1980).

Gelfand, I. M., Fomin, S. V., Silverman, R. A. (tr), 2000, *Calculus of Variations* (Dover Publications, Mineola, NY, 2000).

Giaquinta, M., Modica, G., Souchek, J., 1998, *Cartesian Coordinates in the Calculus of Variations* Volumes I and II (Springer-Verlag, New York, 1998).

Giaquinta, M., Hildebrandt, S., 1996, *Calculus of Variations* Volumes I and II (Springer-Verlag, New York, 1996).

Gradshteyn, I. S. and Ryzhik, I. M., 1965, *Table of Integrals, Series, and Products* (Academic Press, New York, 1965).

Heitler, W., 1954, *The Quantum Theory of Radiation* (Claendon Press, Oxford, UK, 1954).

Huang, Kerson, 1992, *Quarks, Leptons & Gauge Fields 2nd Edition* (World Scientific Publishing Company, Singapore, 1992).

Jost, J., Li-Jost, X., 1998, *Calculus of Variations* (Cambridge University Press, New York, 1998).

Kaku, Michio, 1993, *Quantum Field Theory*, (Oxford University Press, New York, 1993).

Kirk, G. S. and Raven, J. E., 1962, *The Presocratic Philosophers* (Cambridge University Press, New York, 1962).

Landau, L. D. and Lifshitz, E. M., 1987, *Fluid Mechanics 2nd Edition*, (Pergamon Press, Elmsford, NY, 1987).

Misner, C. W., Thorne, K. S., and Wheeler, J. A., 1973, *Gravitation* (W. H. Freeman, New York, 1973).

Rescher, N., 1967, *The Philosophy of Leibniz* (Prentice-Hall, Englewood Cliffs, NJ, 1967).

Rieffel, Eleanor and Polak, Wolfgang, 2014, *Quantum Computing* (MIT Press, Cambridge, MA, 2014).

Riesz, Frigyes and Sz.-Nagy, Béla, 1990, *Functional Analysis* (Dover Publications, New York, 1990).

Sagan, H., 1993, *Introduction to the Calculus of Variations* (Dover Publications, Mineola, NY, 1993).

Sakurai, J. J., 1964, *Invariance Principles and Elementary Particles* (Princeton University Press, Princeton, NJ, 1964).

Sorokin, Pitirim, 1941, *Social and Cultural Dynamics* (Porter Sargent Publishers, Boston, MA, 1941).

Streater, R. F. and Wightman, A. S., 2000, *PCT, Spin, Statistics, and All That* (Princeton University Press, Princeton, NJ 2000).

Weinberg, S., 1972, *Gravitation and Cosmology* (John Wiley and Sons, New York, 1972).

Weinberg, S., 1995, *The Quantum Theory of Fields Volume I* (Cambridge University Press, New York, 1995).

Weinberg, S., 2000, *The Quantum Theory of Fields Volume III Supersymmetry* (Cambridge University Press, New York, 2000).

Weyl, H., 1950, *Space, Time, Matter* (Dover, New York, 1950).

INDEX

About the Author

Stephen Blaha is a well known Physicist and Man of Letters with interests in Science, Society and civilization, the Arts, and Technology. He had an Alfred P. Sloan Foundation scholarship in college. He received his Ph.D. in Physics from Rockefeller University. He has served on the faculties of several major universities. He was also a Member of the Technical Staff at Bell Laboratories, a manager at the Boston Globe Newspaper, a Director at Wang Laboratories, and President of Blaha Software Inc and of Janus Associates Inc. (NH).

Among other achievements he was a co-discoverer of the "r potential" for heavy quark binding developing the first (and still the only demonstrable) non-abelian gauge theory with an "r" potential; first suggested the existence of topological structures in superfluid He-3; first proposed Yang-Mills theories would appear in condensed matter phenomena with non-scalar order parameters; first developed a grammar-based formalism for quantum computers and applied it to elementary particle theories; first developed a new form of quantum field theory without divergences (thus solving a major 60 year old

problem that enabled a unified theory of the Standard Model and Quantum Gravity without divergences to be developed); first developed a formulation of complex General Relativity based on analytic continuation from real space-time; first developed a generalized non-homogeneous Robertson-Walker metric that enabled a quantum theory of the Big Bang to be developed without singularities at t = 0; first generalized Cauchy's theorem and Gauss' theorem to complex, curved multi-dimensional spaces; received Honorable Mention in the Gravity Research Foundation Essay Competition in 1978; first developed a physically acceptable theory of faster-than-light particles; first derived a composition of extrema method in the Calculus of Variations; first quantitatively suggested that inflationary periods in the history of the universe were not needed; first proved Gödel's Theorem implies Nature must be quantum; provided a new alternative to the Higgs Mechanism, and Higgs particles, to generate masses; first showed how to resolve logical paradoxes including Gödel's Undecidability Theorem by developing Operator Logic and Quantum Operator Logic; first developed a quantitative harmonic oscillator-like model of the life cycle, and interactions, of civilizations; first showed how equations describing superorganisms also apply to civilizations. A recent book shows his theory applies successfully to the past 14 years of history and to *new* archaeological data on Andean and Mayan civilizations as well as Early Anatolian and Egyptian civilizations.

He first developed an axiomatic derivation of the form of The Standard Model from geometry – space-time properties – The Unified SuperStandard Model. It unifies all the known forces of Nature. It also has a Dark Matter sector that includes a Dark ElectroWeak sector with Dark doublets and Dark gauge interactions. It uses quantum coordinates to remove infinities that crop up in most interacting quantum field theories and additionally to remove the infinities that appear in the Big Bang and generate inflationary growth of the universe. It shows gravity has a MOND-like form without sacrificing Newton's Laws. It relates the interactions of the MOND-like sector of gravity with the r-potential of Quark Confinement. The axioms of the theory lead to the question of their origin. We suggest in the preceding edition of this book it can be attributed to an entity with God-like properties. We explore these properties in "God Theory" and show they predict that the Cosmos exists forever although individual universes (or incarnations of our universe) "come and go." Several other important results emerge from God Theory such a a functionally triune God. The Unified SuperStandard Theory has many other important parts described in the Current Edition of *The Unified Superstandard Theory* and expanded in subsequent volumes.

Blaha has had a major impact on a succession of elementary particle theories: his Ph.D. thesis (1970), and papers, showed that quantum field theory calculations to all orders in ladder approximations could not give scaling deep inelastic electron-nucleon scattering. He later showed the eigenvalue equation for the fine structure constant α in Johnson-Baker-Willey QED had a zero at α = 1 not 1/137 by solving the Schwinger-Dyson equations to all orders in an approximation that agreed with exact results to 4th order in α thus ending interest in this theory. In 1979 at Prof. Ken Johnson's (MIT) suggestion he calculated the proton-neutron mass difference in the MIT bag model and found the result had the wrong sign reducing interest in the bag model. These results all appear in Physical Review papers. In the 2000's he repeatedly pointed out the shortcomings of SuperString theory and showed that The Standard Model's form could be derived from space-time geometry by an extension of Lorentz transformations to faster than light transformations. This deeper space-time basis greatly increases the possibility that it is part of THE fundamental theory.Recently, Blaha showed that the Weak interactions differed significantly from the Strong, electromagnetic and gravitation interactions in important respects while these interactions had similar features, and suggested that ElectroWeak theory, which is essentially a glued union of the Weak interactions and Electromagnetism, possibly modulo unknown Higgs particle features, be replaced by a unified theory of the other interactions combined with a stand-alone Weak interaction theory. Blaha also showed that, if Charmonium calculations are taken seriously, the Strong interaction coupling constant is only a factor of five larger than the electromagnetic coupling constant, and thus Strong interaction perturbation theory would make sense and yield physically meaningful results.

In graduate school (1965-71) he wrote substantial papers in elementary particles and group theory: The Inelastic E- P Structure Functions in a Gluon Model. Phys. Lett. B40:501-502,1972; Deep-Inelastic E-P Structure Functions In A Ladder Model With Spin 1/2 Nucleons, Phys.Rev. D3:510-523,1971; Continuum Contributions To The Pion Radius, Phys. Rev. 178:2167-2169,1969; Character Analysis of U(N) and SU(N), J. Math. Phys. 10, 2156 (1969); and The Calculation of the Irreducible Characters of the Symmetric Group in Terms of the Compound Characters, (Published as Blaha's Lemma in D. E. Knuth's book: *The Art of Computer Programming Vols. 1 – 4*).

In the early 1980's Blaha was also a pioneer in the development of UNIX for financial, scientific and Internet applications: benchmarked UNIX versions showing that block size was critical for UNIX performance, developing financial modeling software, starting database benchmarking comparison studies, developing Internet-like UNIX networking (1982) and developing a hybrid shell programming technique (1982) that was a precursor to the PERL programming language. He was also the manager of the AT&T ten-year future products development database. His work helped lead to commercial UNIX on computers such as Sun Micros, IBM AIX minis, and Apple computers.

In the 1980's he pioneered the development of PC Desktop Publishing on laser printers. and was nominated for three "Awards for Technical Excellence" in 1987 by PC Magazine for PC software products that he designed and developed.

Recently he has developed a theory of Megaverses – actual universes of which our universe is one – with quantum particle-like properties based on the Wheeler-DeWitt equation of Quantum Gravity. He has developed a theory of a baryonic force, which had been conjectured many years ago, and estimated the strength of the force based on discrepancies in measurements of the gravitational constant G. This force, operative in D-dimensinal space, can be used to escape from our universe in "uniships" which are the equivalent of the faster-than-light starships proposed in the author's earlier books. Thus travel to other universes, as well as to other stars is possible.

Blaha also considered the complexified Wheeler-DeWitt equation and showed that its limitation to real-valued coordinates and metrics generated a Cosmological Constant in the Einstein equations.

The author has also recently written a series of books on the serious problems of the United States and their solution as well as a book on the decline of Mankind that will follow from current social and genetic trends in Mankind.

In the past twelve years Dr. Blaha has written over 40 books on a wide range of topics. Some recent major works are: *From Asynchronous Logic to The Standard Model to Superflight to the Stars, All the Universe!, SuperCivilizations: Civilizations as Superorganisms, America's Future: an Islamic Surge, ISIS, al Qaeda, World Epidemics, Ukraine, Russia-China Pact, US Leadership Crisis,The Rises and Falls of Man – Destiny – 3000 AD: New Support for a Superorganism MACRO-THEORY of CIVILIZATIONS From CURRENT WORLD TRENDS and NEW Peruvian, Pre-Mayan, Mayan, Anatolian, and Early Egyptian Data, with a Projection to 3000 AD*, and *Mankind in Decline: Genetic Disasters, Human-Animal Hybrids, Overpopulation, Pollution, Global Warming, Food and Water Shortages, Desertification, Poverty, Rising Violence, Genocide, Epidemics, Wars, Leadership Failure.*

He has taught approximately 4,000 students in undergraduate, graduate, and postgraduate corporate education courses primarily in major universities, and large companies and government agencies.

www.ingramcontent.com/pod-product-compliance
Lightning Source LLC
Chambersburg PA
CBHW082006190326
41458CB00010B/3095